T0189340

Domain Adaptation in Computer Vision with Deep Learning

Hemanth Venkateswara • Sethuraman Panchanathan
Editors

Domain Adaptation in Computer Vision with Deep Learning

 Springer

Editors
Hemanth Venkateswara
Center for Cognitive Ubiquitous Computing
(CUbiC)
School of Computing Informatics and
Decision Systems Engineering
Arizona State University
Tempe, AZ, USA

Sethuraman Panchanathan
Center for Cognitive Ubiquitous Computing
(CUbiC)
School of Computing Informatics and
Decision Systems Engineering
Arizona State University
Tempe, AZ, USA

ISBN 978-3-030-45531-6   ISBN 978-3-030-45529-3   (eBook)
https://doi.org/10.1007/978-3-030-45529-3

This Springer imprint is published by the registered company Springer Nature Switzerland AG
The registered company address is: Gewerbestrasse 11, 6330 Cham, Switzerland

Preface

The focus of this book on Domain Adaptation in Computer Vision With Deep Learning is to serve as a one-stop shop for deep learning-based computer vision research in domain adaptation. The book is also meant to be a concise guide for navigating the vast amount of research in this area. The book is organized into four parts that provide a summary of research in domain adaptation. It begins with an introduction to domain adaptation and a survey of non-deep learning-based research in the first part. In Parts **??** and **??**, the book discusses feature alignment and image alignment techniques for domain adaptation. Part **??** of the book outlines novel approaches detailing the future of research in domain adaptation.

A diverse set of experts were invited to contribute comprehensive and complementary perspectives. The editors thank the contributing authors for sharing their perspectives. The editors also acknowledge the funding support of Arizona State University and the National Science Foundation (Grant No. 1828010), which made this book project possible.

Tempe, AZ, USA
January 2020

Hemanth Venkateswara
Sethuraman Panchanathan

Contents

Contributors

Amir Atapour-Abarghouei Department of Computer Science, Durham University, Durham, UK

Vineeth N. Balasubramanian Indian Institute of Technology, Hyderabad, India

Toby P. Breckon Departments of Engineering and Computer Science, Durham University, Durham, UK

Zhangjie Cao School of Software, Tsinghua University, Beijing, China

Shayok Chakraborty Florida State University, Tallahassee, FL, USA

Qingchao Chen University of Oxford, Oxford, UK

Ziliang Chen Sun Yat-sen University, Guangzhou, People's Republic of China

Xilin Chen Key Lab of Intelligent Information Processing of Chinese Academy of Sciences, Institute of Computing Technology, Chinese Academy of Sciences, Beijing, China

Kevin Chetty University College London, London, UK

Jose Eusebio Axosoft Scottsdale, AZ, USA

Tatsuya Harada The University of Tokyo and RIKEN, Tokyo, Japan

Xiaodong He JD AI Research, Beijing, China

Lanqing Hu Key Lab of Intelligent Information Processing of Chinese Academy of Sciences, Institute of Computing Technology, Chinese Academy of Sciences, Beijing, China

Michael I. Jordan University of California, Berkeley, CA, USA

Meina Kan Key Lab of Intelligent Information Processing of Chinese Academy of Sciences, Institute of Computing Technology, Chinese Academy of Sciences, Beijing, China

Kyungnam Kim HRL Laboratories, LLC, Malibu, CA, USA

Soheil Kolouri HRL Laboratories, LLC, Malibu, CA, USA

David Kriegman Department of Computer Science & Engineering, University of California, San Diego (UCSD), La Jolla, CA, USA

Narayanan Chatapuram Krishnan Indian Institute of Technology Ropar, Rupnagar, Punjab, India

Kuang-Huei Lee Microsoft AI and Research, Redmond, WA, USA
Google Brain, San Francisco, CA, USA

Liang Lin Sun Yat-sen University, Guangzhou, People's Republic of China

Yang Liu University of Oxford, Oxford, UK

Mingsheng Long School of Software, Tsinghua University, Beijing, China

Orchid Majumder AWS AI, Seattle, WA, USA

Zak Murez HRL Laboratories, LLC, Malibu, CA, USA

Bhadrinath Nagabandi Arizona State University, Tempe, AZ, USA

Arghya Pal Indian Institute of Technology, Hyderabad, India

Sethuraman Panchanathan Center for Cognitive Ubiquitous Computing (CUbiC), School of Computing Informatics and Decision Systems Engineering, Arizona State University, Tempe, AZ, USA

Raghavendran Ramakrishnan Arizona State University, Tempe, AZ, USA

Ravi Ramamoorthi Department of Computer Science & Engineering, University of California, San Diego (UCSD), La Jolla, CA, USA

Kuniaki Saito The University of Tokyo, Tokyo, Japan

Shiguang Shan Key Lab of Intelligent Information Processing of Chinese Academy of Sciences, Institute of Computing Technology, Chinese Academy of Sciences, Beijing, China

Sanatan Sukhija Indian Institute of Technology Ropar, Rupnagar, Punjab, India

Gurumurthy Swaminathan AWS AI, Seattle, WA, USA

Yoshitaka Ushiku The University of Tokyo, Tokyo, Japan

Ragav Venkatesan AWS AI, Seattle, WA, USA

Hemanth Venkateswara Center for Cognitive Ubiquitous Computing (CUbiC), School of Computing Informatics and Decision Systems Engineering, Arizona State University, Tempe, AZ, USA

Zhaowen Wang Adobe Research, San Jose, CA, USA

Jianmin Wang School of Software, Tsinghua University, Beijing, China

Ian Wassell University of Cambridge, Cambridge, UK

Xiang Xu AWS AI, Seattle, WA, USA

Shohei Yamamoto The University of Tokyo, Tokyo, Japan

Linjun Yang Facebook, Seattle, WA, USA

Kaichao You School of Software, Tsinghua University, Beijing, China

Lei Zhang Microsoft Research, Redmond, WA, USA

Xiong Zhou AWS AI, Seattle, WA, USA

Part I
Introduction

Chapter 1
Introduction to Domain Adaptation

Hemanth Venkateswara and Sethuraman Panchanathan

1.1 Introduction

In a Harvard Business Review article in 2016, leading computer science researcher Andrew Ng, compared artificial intelligence to electricity, saying, "*A hundred years ago electricity transformed countless industries; 20 years ago the internet did, too. Artificial intelligence is about to do the same,*" [57]. Research in artificial intelligence has been accelerating in the recent past. All areas of scientific research from social sciences to neuroscience are applying the tools of artificial intelligence to find new insights and further their research. Powered by the success of deep learning, artificial intelligence is able to rival human performance in solving challenging niche problems, like playing Go and chess. However, replicating and eventually outperforming general human intelligence is the utopian goal of researchers in artificial intelligence.

Human intelligence is a competitive benchmark for modern artificial intelligence systems. One of the hallmarks of human intelligence is it's ability to seamlessly adapt to novel environments, and its capacity to build upon pre-existing concepts when learning new tasks. For e.g., once a person learns to ride a bicycle, they can easily transition to riding a motorcycle, without the need to learn motorcycle riding from scratch. On the other hand, machine learning systems have to be explicitly trained from scratch to solve new tasks. In order to emulate human intelligence we will need to build artificial intelligence systems that can engage in continual learning by building upon pre-existing knowledge concepts and transferring that knowledge to adapt to new environments. While simulating human intelligence in machines may not be a pressing need, building machine learning systems efficiently, definitely

H. Venkateswara (✉) · S. Panchanathan
Center for Cognitive Ubiquitous Computing (CUbiC), School of Computing Informatics and Decision Systems Engineering, Arizona State University, Tempe, AZ, USA
e-mail: hemanthv@asu.edu; panch@asu.edu

is. Machine learning systems can be built efficiently with limited human supervision if we can model knowledge and its transfer. This area of research in machine learning is called transfer learning and it deals with building machine learning models that can transfer knowledge. In this chapter we will introduce different types of knowledge transfer in machine learning including domain adaptation. The rest of the chapters discuss different approaches to solving domain adaptation using deep learning.

1.2   Paradigms of Learning

We will begin with defining different paradigms of learning. The traditional paradigms of machine learning are *unsupervised learning, supervised learning* and *semi supervised learning*. In recent years a new paradigm of machine learning has emerged—*self supervised learning*.

Unsupervised Learning   In the unsupervised learning paradigm data is presented as a set of n examples, $\mathcal{X} = \{x_1, \ldots, x_n\}$, where $x_i \in \mathcal{X}$ for all $i \in [1, \ldots, n]$. $\mathcal{X}$ represents the feature space of the data. For e.g., $\mathcal{X}$ could be a subset of the Euclidean space of d-dimensions, $\mathbb{R}^d$, or a space of normalized color images of dimensions $M \times N \times C$, $[0, 1]^{M \times N \times C}$ and so on. The dataset $\mathcal{X}$ is drawn from a probability distribution $P(X)$, which is unknown. Here, X is a random variable and $P(X = x)$ is the probability of data instance x. Traditionally, unsupervised learning involved estimating structure and patterns in the data like clustering, quantile estimation, dimensionality reduction and outlier detection [11]. In some cases the goal of unsupervised learning is to estimate the probability distribution $P(X)$ that generated the samples in $\mathcal{X}$. The probability distribution $P(.)$ is used to generate new samples and this class of unsupervised learning algorithms are referred to as *generative models*. Popular approaches to generative modeling are autoregressive models [86], generative adversarial networks [27] and variational autoencoders [41].

Supervised Learning   In the supervised learning paradigm, data is presented as a set of n tuples, $\mathcal{X}_l = \{x_i, y_i\}_{i=1}^n$, where $x_i \in \mathcal{X}$ and $y_i \in \mathcal{Y}$ are called the labels for all $i \in [1, \ldots, n]$. The labels space $\mathcal{Y}$ could be discrete, with $y_i \in [1, 2, \ldots, C]$ for problems in classification. In the case of regression, the label space $\mathcal{Y}$ could be real with $y_i \in \mathbb{R}$. It could also belong to the same space $\mathcal{X}$ as in the case of image segmentation. These problems are referred to as structured prediction [58]. In supervised learning, data is sampled from a joint distribution $P(X, Y)$. Here, X and Y are random variables and $P(X = x, Y = y)$ is joint probability of sample (x, y). The goal of supervised learning is to estimate a mapping function $f : \mathcal{X} \to \mathcal{Y}$ using the training data $\mathcal{X}_l$. There are two standard approaches to estimating the mapping function. *Generative* approaches learn to predict the posterior probability $P(Y|X)$ when they have an estimate for the marginal distribution $P(X)$, the conditional distribution $P(X|Y)$ and the prior $P(Y)$, using the Bayes formula,

$P(Y|X) = \frac{P(X|Y)P(Y)}{P(X)}$. Examples of generative approaches are Gaussian Mixture Models, Hidden Markov Models and Naïve Bayes. The second is the *discriminative* approach. In the discriminative approach, the posterior distribution is directly modeled without the need to estimate the conditional and prior distributions. Examples of the discriminative approach Support Vector Machine, Logistic Regression, Neural Networks and Random Forests.

Semi Supervised Learning In the semi supervised learning paradigm data is presented as two datasets, $\mathcal{X}_l = \{x_i, y_i\}_{i=1}^{n_l}$ and $\mathcal{X}_u = \{x_i\}_{i=1}^{n_u}$, where $\mathcal{X}_l$ is the labeled dataset and $\mathcal{X}_u$ is the unlabeled dataset. The labeled data $\mathcal{X}_l$ is drawn from a joint distribution $P(X, Y)$ and the unlabeled data is drawn from the marginal distribution $P(X)$. The goal in semi supervised learning is to estimate the labels for $\mathcal{X}_u$. There are two approaches to solving the semi supervised learning problem; *transductive* and *inductive*. A transductive model estimates the labels only for the unlabeled samples $\mathcal{X}_u$. An example of the transductive approach is the Transductive SVM [40]. On the other hand, the inductive model learns the mapping function $f : \mathcal{X} \rightarrow \mathcal{Y}$ based on the labeled example pairs in $\mathcal{X}_l$ and unlabeled examples $\mathcal{X}_u$. The challenge in this approach is that n_l is very small and insufficient to train a robust classification model. However, the number of unlabeled samples is large with $n_u \gg n_l$ and they can used to estimate $f(.)$. Modern deep learning based approaches make two assumptions for semi supervised learning—*smoothness* and *cluster*. The smoothness assumption ensures smoothly varying label spaces or decision boundaries. In other words, data points in a neighborhood tend to have the same label. The smoothness assumption is implemented using consistency regularization [81]. The cluster assumption ensures that decision boundaries pass through regions of low data denisty. The cluster assumption is implemented using entropy regularization [30]. In the following sections we will discuss how domain adaptation can be reduced to semi supervised learning.

Self Supervised Learning In recent years self supervised learning techniques have been developed in natural language processing and computer vision. Self supervised learning is an extension of unsupervised learning where we have unlabeled data $\mathcal{X} = \{x_1, \ldots, x_n\}$, where $x_i \in \mathcal{X}$ for all $i \in [1, \ldots, n]$ and the data does not have labels. $\mathcal{X}$ represents the feature space of the data. But unlike unsupervised learning, the unlabeled data is modified to generate pseudo labeled data to mimic a supervised learning set up. In order to qualify as a self supervised learning approach, the pseudo labels should be generated automatically from the unlabeled data without the need for human annotation. These pseudo labels are then used to train a pretext task in a supervised manner. For example, in the BERT model for language, a pretext task can be to predict the next word (or the previous word) in a sentence [16]. In vision problems, the pretest task could be to determine the rotation of a pre-rotated image [25], or determine the relative position between two square patches cropped from an image grid [17]. Other pre-tasks can be the coloring of a gray scaled image [100], or the frame order prediction in a frame-shuffled video [54]. A survey of self supervised learning techniques is discussed in [39]. Deep neural networks trained on such pretext tasks can be viewed as very good feature extractors that have captured the essence of image semantics. These pre-trained deep neural networks are then fine-tuned to learn the original task with a few labeled samples. These techniques have also been applied towards domain adaptation [38].

1.3 Types of Knowledge Transfer

The paradigms of machine learning discussed in the previous section can be considered as *static* models. These models are developed with the datasets present at the time of training and they model the probabilities of the training dataset. These models do not adapt to changes in the data distribution. In other words, if the test data is sampled from a different distribution, these models do not guarantee optimal performance. For e.g., consider a facial expression recognition algorithm trained on images consisting of only women. During evaluation, if the image presented is that of a male subject, the algorithm may perform suboptimally because the data comes from a different distribution—men may have more rugged features and even facial hair compared to women. *Transfer learning* is the branch of machine learning where models trained on data from one distribution are adapted to data from other distributions. Transfer learning is perhaps inspired by the way humans learn by adapting. Humans learn by building upon concepts and rarely learn concepts in isolation [93].

In this section we will discuss different kinds of knowledge transfer in machine learning. We lay the groundwork with a few definitions and notations as per Pan et al. [61]. We define a domain $\mathcal{D}$ to consist of two components, a feature space $\mathcal{X}$ and a marginal probability distribution $P(X)$ for that feature space, with $\mathcal{X} = \{x_1, \ldots, x_n\} \subset \mathcal{X}$ a set of samples from the feature space sampled using $P(X)$. If we consider the example of audio transcription, audio samples from multiple subjects can be viewed as different domains. The voice, diction and vocabulary of a particular subject can be considered as the feature space $\mathcal{X}$ and the samples $\mathcal{X} = \{x_1, \ldots, x_n\}$ can be viewed as the set of audio samples (utterances) by the subject sampled using a probability distribution $P(X)$. In other definitions, a domain $\mathcal{D}$ can also be represented as a space of features, labels and a joint probability distribution, $\{(\mathcal{X} \times \mathcal{Y}), P(X, Y)\}$. The difference in domains, also termed as domain disparity, occurs when either the probability distributions are different (for e.g., casual conversation vs. reading) or when the feature spaces are different (for e.g., different subjects) or both. If $\mathcal{D} = \{\mathcal{X}, P(X)\}$ refers to a domain, a task $\mathcal{T}$ is defined as $\mathcal{T} = \{\mathcal{Y}, f(.)\}$. Here, $\mathcal{Y}$ is the label space and $f : \mathcal{X} \rightarrow \mathcal{Y}$ is a function mapping data points in $\mathcal{X}$ to the labels in $\mathcal{Y}$. In the supervised learning paradigm, the function $f(.)$ is determined using labeled data points (x, y) with $x \in \mathcal{X}$ and $y \in \mathcal{Y}$. The function $f(x)$ can also be viewed as the posterior probability $P(y|x)$. In traditional transfer learning, a pair of domains is usually considered, the *source* and the *target*, to model transfer of knowledge. With a slight abuse of notation, we represent a source dataset as a collection of data points

$\mathcal{D}_s = \{(\boldsymbol{x}_1^s, y_1^s), \ldots, (\boldsymbol{x}_{n_s}^s, y_{n_s}^s)\}$, with $\boldsymbol{x}_i^s \in \mathcal{X}_S$ and $y_i \in \mathcal{Y}_S$. Likewise, we represent a target dataset as $\mathcal{D}_t = \{(\boldsymbol{x}_1^t, y_1^t), \ldots, (\boldsymbol{x}_{n_t}^t, y_{n_t}^t)\}$, with $\boldsymbol{x}_i^t \in \mathcal{X}_T$ and $y_i \in \mathcal{Y}_T$. We begin our discussion on transfer learning with the following definition.

Definition (Transfer Learning [61]) Given a source domain $\mathcal{D}_S$ and a source learning task $\mathcal{T}_S$, a target domain $\mathcal{D}_T$ and a target learning task $\mathcal{T}_T$, transfer learning aims to improve the target predictive function $f_T(.)$ using $\mathcal{D}_S$ and $\mathcal{T}_S$, where $\mathcal{D}_S \neq \mathcal{D}_T$, or $\mathcal{T}_S \neq \mathcal{T}_T$. □

Two domains are considered to be different ($\mathcal{D}_S \neq \mathcal{D}_T$), if either the feature spaces are different ($\mathcal{X}_S \neq \mathcal{X}_T$, as in audio from two different subjects) or the probability distributions are not the same ($P_S(X) \neq P_T(X)$, as in casual talk vs. reading a report) or both. Likewise, two tasks are considered to be different ($\mathcal{T}_S \neq \mathcal{T}_T$, where $\mathcal{T}_S = \{\mathcal{Y}_S, P_S(Y|X)\}$ and $\mathcal{T}_T = \{\mathcal{Y}_T, P_T(Y|X)\}$), when either the label spaces are not the same ($\mathcal{Y}_S \neq \mathcal{Y}_T$, as in audio transcription vs. sentiment analysis) or when the posterior distributions are not the same ($P_S(Y|X) \neq P_T(Y|X)$, the case where the source and target have different label spaces). These differences in domains and tasks give rise to learning scenarios that are addressed by transfer learning. Transfer learning models extract knowledge from one or more tasks, domains or data distributions and apply that knowledge to develop effective hypotheses for a different domain, task or distribution [6]. This chapter is partially based on earlier work from [90]. In the following we outline and differentiate popular problems in machine learning that incorporate elements of knowledge transfer, viz., *multitask learning, self-taught learning, sample selection bias, lifelong machine learning, zero-shot learning* and *domain adaptation*.

1.3.1 Multitask Learning (MTL)

Under this setting we consider K tasks $\mathcal{T} = \{\mathcal{T}_1, \mathcal{T}_2, \ldots, \mathcal{T}_K\}$ where the data for each task is sampled from K different domains $\mathcal{D} = \{\mathcal{D}_1, \mathcal{D}_2, \ldots, \mathcal{D}_K\}$ respectively. Each of the individual tasks is supervised with labeled data. However, we are also faced with the constraint that it may not be possible to estimate a reliable empirical probability distribution $\hat{P}_k(X, Y)$ for the kth domain using only data from the kth domain $\mathcal{D}_k = \{x_k^i, y_k^i\}_{i=1}^{n_k}$, $x_k^i \in \mathcal{X}_k$ and $y_k^i \in \mathcal{Y}_k$. In the multitask learning environment, the tasks are different whereas the domains may be the same or related. Multitask learning algorithms attempt to simultaneously estimate the empirical distributions $\hat{P}_k(X, Y)$ for all the K tasks by using data from all the domains, $\mathcal{D} = \{\mathcal{D}_1, \mathcal{D}_2, \ldots, \mathcal{D}_K\}$. The idea in multitask learning is that by simultaneously learning all the tasks in a single model we enable the model to generalize better across each of the tasks. Multitask learning algorithms improve generalization by leveraging domain-specific information across all the tasks [10].

Deep neural network based multitask learning is usually implemented either with hard parameter sharing [3], where all tasks share the same parameters of a base network and diversify in the last layer or with soft parameter sharing [96], where the parameters of multiple networks are constrained to be similar using ℓ_2 norm. In other variations of parameter sharing we have coupled networks with cross-stitches which allow the model to choose task specific parameters from either network [53]. Sluice networks are the most generalized form of parameter sharing that enables sharing at multiple levels and across subspaces [70]. A survey of multitask learning procedures from the pre-deep learning era and modern day deep learning approaches is provided in [83]and [69] respectively.

1.3.2 Self-Taught Learning

This learning paradigm is inspired by how humans learn from unlabeled data in an unsupervised manner [67]. In this paradigm of learning, knowledge transfer occurs from unrelated domains using learned representations. Provided with unlabeled data, $\{x_u^1, \ldots, x_u^k\}$ where $x_u^i \in \mathbb{R}^d$, the self-taught learning framework learns a set of K basis vectors which are then used as basis to represent the target data features. Specifically,

$$\min \sum_i ||x_u^i - \sum_j^K a_i^j b_j||^2 + \beta||a_i||_1$$

$$\text{s.t. } ||b_j|| \leq 1, \forall j \in 1, \ldots K \tag{1.1}$$

where, $\{b_1, \ldots, b_K\}$ are the set of basis vectors which are learned using unlabeled data with $b_i \in \mathbb{R}^d$. The corresponding sparse representation for input data x_u^i is $a_i = \{a_i^1, \ldots, a_i^K\}$ where a_i^j corresponds to the basis vector b_j. When the same set of basis vectors $\{b_1, \ldots, b_K\}$ is used to represent the labeled target data, we consider transfer of knowledge to have occurred. Some of the popular computer vision and machine learning techniques that implement self-taught learning are [4, 43, 50, 97].

In the era of deep learning, self-taught learning using a set of basis vectors, as described above, is not quite relevant. However, the principle of transferring knowledge using neural networks trained on unrelated datasets is quite relevant and is commonly referred to as transfer learning. In this form of naïve transfer learning, a neural network trained on a generic dataset is fine-tuned to solve a new task. For e.g., most modern domain adaptation techniques use pre-trained models like AlexNet and ResNet and fine-tune them to their specific classification tasks [47, 98]. This is related to self-taught learning where feature vectors from random data form basis vectors for the target data.

1.3.3 Sample Selection Bias

John Heckman in 1979, introduced the concept of sample selection bias in his Nobel prize winning work [32]. Sample selection bias occurs when a distribution of a sample of data does not reflect the true distribution of population it is meant to reflect. Consider an example where a financial bank wants to create a model profile of people who default on their loans in order to deny them future loans. Using the profile of defaulters in its records it builds a model. However, the bank only has a few records of defaulters and they do not reflect the distribution of loan defaulters in the general public. The bank's model is therefore offset or biased due to the inadequacy of its sample. This offset is termed as sample selection bias.

In this mode of transfer learning there exists a sample labeled dataset $\mathcal{D} = \{x_i, y_i\}_{i=1}^n$. The goal is to determine the joint distribution $\hat{P}(X, Y)$ which is a true approximation to the joint distribution $P(X, Y)$ of the population using $\mathcal{D}$. Since $\mathcal{D}$ is only a tiny subset of the entire population, the approximate $\hat{P}(X, Y)$ is not the same as $P(X, Y)$ [6]. This may be because the tiny subset $\mathcal{D}$ may lead to an incorrect estimation of the true marginal distribution $P(X)$ with $\hat{P}(X) \neq P(X)$. It could also be due to an incorrect estimate of the class prior with $\hat{P}(Y) \neq P(Y)$ which then leads to an incorrect estimate of the class conditional $\hat{P}(Y|X) \neq P(Y|X)$. When there is a bias in the marginal distribution ($\hat{P}(X) \neq P(X)$) as well as the conditional distributions ($\hat{P}(Y|X) \neq P(Y|X)$), this is referred to as sample selection bias and the following works address knowledge transfer under these settings [18, 37, 99]. If there is a distribution difference in the prior probabilities ($\hat{P}(X) \neq P(X)$) and the conditional distributions can be considered to be equal ($\hat{P}(Y|X) \approx P(Y|X)$) this situation is referred to as covariate shift and the following papers discuss knowledge transfer under covariate shift [5, 31, 66, 74].

1.3.4 Lifelong Machine Learning (LML)

The paradigm of lifelong learning was introduced by Thrun in the seminal work [82]. Transfer in lifelong learning is outlined as follows. A machine learning model that was trained for tasks $\{\mathcal{T}_1, \mathcal{T}_2, \ldots, \mathcal{T}_K\}$ is updated to learn task $\mathcal{T}_{K+1}$ with data $\mathcal{D}_{K+1}$. The argument for LML is that learning the $K + 1^{th}$ task is easier since the model has already learned tasks $\{\mathcal{T}_1, \mathcal{T}_2, \ldots, \mathcal{T}_K\}$. The key ideas behind lifelong learning are: (1) a continual learning process, (2) accumulation of knowledge, and (3) transfer of previous knowledge to help with future learning [20].

Lifelong machine learning is different from multitask learning because it retains knowledge from previous tasks and transfers that knowledge to learn new tasks whereas multitask learning learns all the tasks at once. It is also different from domain adaptation where the transfer of knowledge is usually to learn a single target task. Lifelong learning is closely related to the paradigm of *incremental learning* or *continual learning*, where a model is trained to learn a new task while not forgetting previously learned tasks [15]. When learning a new task, some incremental learning models depend upon data from previous tasks, in order to avoid the problem of *catastrophic forgetting* [52]. In other approaches to incremental learning, succinct data representations, also termed as exemplars, are used to model data from previous tasks. These exemplars are recalled when updating the classifier with a new task [68]. When no data is used from previous tasks and the model is updated only using data from the new task, it is treated as an uncompromising form of incremental learning [45, 87]. A survey of continual learning approaches can be found in [15, 65].

1.3.5   Zero-Shot and Few-Shot Learning

This paradigm of learning can be viewed as an extreme case of knowledge transfer [28]. These approaches attempt to learn data categories from minimal data. The key idea is the ability to transfer knowledge of previously learned categories to learn new categories. In few-shot (or one-shot) learning, the model is trained to recognize a new category using only a few (or no) labeled examples [21]. In other words, the model is trained to estimate decision boundaries between categories using just a few samples. In the case of one-shot learning, a single labeled example from a new category is used to determine the mean of a cluster around which other unseen examples of the same category will cluster.

Zero-shot learning models recognize new categories without having seen any examples of the new category during training. In zero-data learning [42] and zero-shot learning [60, 75] knowledge transfer occurs from training data that is not completely related to the new categories of interest. For example, a model that has been trained to recognize wild animals like zebra, tiger, lion etc. (seen categories), is provided with a description of a new category (unseen category), for e.g., 'horse'. The model is able to relate a horse to the seen categories and recognize it. Zero-shot learning models associate the textual descriptions between the categories to learn the correlations between the categories. These models utilize the modalities of language and vision to solve the zero-shot learning challenge. Modern deep learning based generative zero-shot learning models hallucinate image features of unseen categories based on their textual descriptions [103]. These hallucinations are used to train the classifier model to distinguish between unseen categories [19].

1.3.6   Domain Adaptation

In this example of transfer learning knowledge transfer occurs between two domains, *source* and *target*. Even though there are examples of multi-source domain adaptation as in [12, 51, 79], knowledge transfer between two domains is the popular setting. The source domain $\mathcal{D}_S$ is different from the target domain $\mathcal{D}_T$, and the goal of domain adaptation is to solve a common task $\mathcal{T} = \{\mathcal{Y}, f(.)\}$. For example, in a digit recognition task, the source domain could contain labeled images of digits against a black background and the target domain could consist of unlabeled images of digits against a colored, noisy and cluttered background. Both the domains have the same set of digit categories. The difference between the source and target domain is modeled as the difference between their joint probability distributions $P_S(X, Y) \neq P_T(X, Y)$. In the standard domain adaptation problem we have a large number of source data points and there are no labeled data (or few samples, if any) in the target dataset. Since the target data has no labeled samples or very few labeled samples, it is difficult to estimate the joint distribution $\hat{P}_T(X, Y)$. The goal of domain adaptation is to approximate $\hat{P}_T(X, Y)$ using the source data distribution estimation $\hat{P}_S(X, Y)$, which is made possible since the two domains are 'correlated'. Often, this correlation is modeled under the covariate shift assumption where, $P_S(X) \neq P_T(X)$ and $P_S(Y|X) \approx P_T(Y|X)$. The goal of the domain adaptation problem is to predict the target labels with minimum expected error. For a hypothesis $h(x)$, the expected error on the target label prediction is given by,[1]

$$\epsilon_T(h) = \mathbb{E}_{p_T(X=x)}\mathbb{E}_{p_{(Y=y|X=x)}}\left(h(x) \neq y\right)$$
$$= \sum_x p_T(x)\mathbb{E}_{p_{(y|x)}}\left(h(x) \neq y\right)$$

Multiplying and dividing by $p_S(x)$

$$= \sum_x \frac{p_S(x)}{p_S(x)} p_T(x)\mathbb{E}_{p_{(y|x)}}\left(h(x) \neq y\right)$$

The target error is equivalent to a weighted source error

with weights given by w,

$$\epsilon_T(h) = \epsilon_S(h, w) = \mathbb{E}_{p_S(x)} \frac{p_T(x)}{p_S(x)} \mathbb{E}_{p_{(y|x)}}\left(h(x) \neq y\right) \qquad (1.2)$$

We have applied the covariate shift assumption with $P_S(Y|X) \approx P_T(Y|X)$ and $P_S(X) \neq P_T(X)$ including the assumption of shared support, i.e. ($P_S(X) = 0$ iff $P_T(X) = 0$). The target prediction error is represented in terms of the labeled source samples where each source data point has a weight $\frac{p_T(x)}{p_S(x)}$. This technique is also used to learn a target classifier and is termed as *instance sampling*. Important examples from the source data are sampled to train a classifier that can also classify target samples. These algorithms estimate the weight using Kernel Mean Matching [31] or Kullback-Leibler divergence [76]. In these instances domain adaptation is

---

[1]http://adaptationtutorial.blitzer.com/.

viewed as a case of covariate shift or sample selection bias [66]. *Feature matching* is another technique, where shared feature representations are estimated to align the features of the source and target [48, 62, 63].

Certain other procedures project the data points into feature subspaces that are common to the source and target datasets. A classifier is then trained in the common feature space using the source data and it is expected to classify the target data points accurately since the domains are aligned [22, 26, 33, 49]. The techniques discussed so far can be termed as *fixed representation* approaches. In these approaches, the feature representations of the source and target images are pre-determined and alignment techniques are applied to reduce the domain distribution difference between the source and target. They can also be termed as *shallow domain adaptation* as opposed to *deep domain adaptation* which are based on deep neural networks. In deep domain adaptation, the features are not pre-determined but extracted using deep neural networks. The process of extraction is used to align the source and target distribution thereby achieving well aligned features. Deep learning based approaches have outperformed fixed representation (shallow domain adaptation) approaches [23, 47, 84] and are the most popular approach to solve the problem of domain adaptation in computer vision. The following chapter provides a classification of shallow domain adaptation approaches. The rest of the chapters discuss novel approaches to deep learning based domain adaptation.

In addition there are other forms of machine learning that can be related to transfer learning. In a temporal setting when the distribution of data changes overtime, it is referred to as *concept drift*. In this setting, knowledge transfer happens by retraining models from the past while also adapting it to data in the present [104]. Knowledge transfer can also happen across modalities as in *multimodal learning*. The human brain is the best example of such transfer where there is a continuous transfer of concepts between multiple modalities of vision, sound and touch. It is evident that there is a latent representation of knowledge and concepts in the human brain that is imprinted and updated through inputs from multiple modalities. This is the reason we can describe what we see or touch or we can imagine (as an image) what we hear and feel with our skin etc. Multimodal processing has been applied in deep learning to transfer knowledge across different modalities to enhance performance [85].

1.4   Forms of Domain Adaptation

This subsection outlines the different forms of domain adaptation. These variants introduce constraints on the standard problem of domain adaptation to align the problem to real-world applications.

1.4.1 Supervised or Semi Supervised Domain Adaptation

The source domain in this setting consists of a labeled dataset $\mathcal{D}_s = \{x_i^s, y_i^s\}_{i=1}^{n_s}$ and the target domain consists of a labeled dataset $\mathcal{D}_t = \{x_i^t, y_i^t\}_{i=1}^{n_t} \cup \{x_i^t\}_{i=n_t+1}^{n_t+n_u}$. The target domain has n_t labeled data points and n_u unlabeled target data points with $n_t \ll n_u$. With a small number of labeled data points (n_t) it is not possible to estimate the joint distribution $P_T(X, Y)$ without the risk of over-fitting. On the other hand, the source dataset has a large number of labeled data points with $n_t \ll n_s$. It can be used to estimate the joint distribution $P_S(X, Y)$. In such cases the source dataset $\mathcal{D}_s$ is used along with $\mathcal{D}_t$ to train a classifier for the target data as in [14, 34, 71] and [88]. This inherently involves aligning the distributions of the source and the target data points. When n_s and n_t are of similar size, the problem can also be viewed as a multi-task learning setup.

1.4.2 Unsupervised Domain Adaptation

In this setting the source domain consists of a labeled dataset $\mathcal{D}_s = \{x_i^s, y_i^s\}_{i=1}^{n_s}$ and the target domain consists of an unlabeled dataset $\mathcal{D}_t = \{x_i^t\}_{i=1}^{n_t}$. The task is to estimate the labels of the target dataset using the labeled source dataset $\mathcal{D}_s$ and the unlabeled target dataset $\mathcal{D}_t$. There is no restriction on the number of source and target samples and generally $n_s \approx n_t$. The source data can be used to estimate the joint distribution $P_S(X, Y)$. The source domain is aligned with the target domain in order to approximate $P_T(X, Y)$ as in [26, 29, 49] and [89]. This is the most standard form of domain adaptation that one encounters in domain adaptation literature.

1.4.3 Unconstrained Label Spaces Domain Adaptation

In the standard form of domain adaptation, the source domain and target domain have identical label spaces, i.e., the source and target data points have the same labels sampled from $\mathcal{Y}$. In an unconstrained setup the source and the target label spaces can be different. With the target data unlabeled the unconstrained label space requirement makes this a more generic and practical setup compared to unsupervised domain adaptation. This form of domain adaptation has become popular in recent years and is more challenging than Closed Set domain adaptation which is the standard unsupervised domain adaptation. There are variants of this approach based on the differences in the overlap of the source label space $\mathcal{Y}_s$ and the target label space $\mathcal{Y}_t$.

1. *Partial Domain Adaptation*: In this form of domain adaptation the source label space is a super set of the target label space, i.e., the source dataset has all the categories found in the target dataset including additional categories, $\mathcal{Y}_t \subset \mathcal{Y}_s$.

This version of domain adaptation was introduced by Cao et al. [8] followed by notable works in recent years, [7, 9, 101].

2. *Openset Domain Adaptation*: In this form of domain adaptation, there is an intersection between the label spaces of the source and the target, i.e., $\mathcal{Y}_t \not\subset \mathcal{Y}_s$, $\mathcal{Y}_s \not\subset \mathcal{Y}_t$ and $\mathcal{Y}_s \cap \mathcal{Y}_t \neq \emptyset$. This version of domain adaptation is inspired by Open Set Recognition [24]. Most notable works in this area include [2, 46, 64, 80]. A variant of Openset domain adaptation assumes the source label space is a subset of the target label space, i.e., $\mathcal{Y}_s \subset \mathcal{Y}_t$ [72]. Universal domain adaptation is another extension of Openset domain adaptation where the model is agnostic to the label space of the target [98]. This is the most generic case which subsumes Closet Set, Partial and Openset domain adaptation paradigms.

Other than these standard paradigms for domain adaptation, there is also multi-source domain adaptation where, as the name indicates, there are multiple source domains and one target domain as in [12, 73, 102] and the multiple source domains are adapted to the target to estimate a classifier for the target. In addition, domain adaptation has been combined with zero-shot learning [92], one-shot learning and [35] few-shot learning [55].

1.5   Book Organization

The book has 13 chapters organized into 4 parts. The organization is meant to provide a guide to understanding and mapping the huge amount of literature in domain adaptation.

- **Part I**: This part introduces the problem of domain adaptation and provides a brief survey of non-deep learning based domain adaptation techniques. It consists of 2 chapters.

  Chapter 1 provides an introduction to domain adaptation. It lays down the definition of transfer learning and outlines the different paradigms of knowledge transfer in machine learning algorithms, including domain adaptation. The chapter discusses different types of domain adaptation. This chapter is partially based on earlier work from [90].

  Chapter 2 gives a survey on shallow domain adaptation. It is called *shallow* because it is a survey of domain adaptation algorithms that do not use deep learning procedures. The chapter also provides a summary of heterogeneous domain adaptation where the feature spaces of the source and the target domains are different in addition to domain disparity. This chapter is partially based on earlier work from [77, 78].

- **Part II**: Deep neural networks extract highly discriminative features, and this part discusses different techniques to align the feature spaces of the source and the target domains to perform domain adaptation. This part consists of 3 chapters.

  Chapter 3 discusses a stochastic neighborhood embedding algorithm that projects source and target features to a domain agnostic latent space and applies a modified Hausdorff distance metric to perform domain adaptation with few labeled samples in the target domain. This chapter is partially based on earlier work from [95].

Chapter 4 proposes a hashing procedure to perform domain adaptation. A deep neural network estimates hash codes for input images and computes hash based similarity to estimate the image category. The deep neural network aligns feature spaces using Maximum Mean Discrepancy and also introduces an unsupervised loss over the target samples. This chapter is partially based on earlier work from [91].

Chapter 5 discusses an adversarial feature alignment technique where a domain discriminator aligns features from the source and the target domain by minimizing an Optimal Transport based Earth Mover's distance. The source labels are also re-weighted to match the target label distribution. This chapter is partially based on earlier work from [13].

- **Part III**: This part discusses adaptation in the pixel space by transforming the input images to reduce domain discrepancy. This part consists of 3 chapters.

  Chapter 6 describes an encoder-decoder network that encodes and decodes images from the source and the target. The encoder projects images to a latent space common to the source and the target domains. In order to align the source and target domains in this latent space, the decoder is trained to reconstruct the encoded images into both the domains. The autoencoder framework is guided by a duplex set of discriminators for each of the domains. Although this is not strictly an image space alignment procedure, image translation guides the encoder to align the features. This chapter is partially based on earlier work from [36].

  Chapter 7 discusses an image translation procedure that first projects images from both the domains to a latent space and uses an adversarial network to align the feature representations. The aligned feature representations are then translated into images from each of the domains, and cyclic losses are used to guide the consistency in translations. The technique is demonstrated by training an image segmentation model upon a synthetic dataset and adapting it to a real-world dataset. This chapter is partially based on earlier work from [56].

  Chapter 8 outlines a domain adaptation procedure based on image style transfer. This chapter discusses the estimation of a depth image from a single RGB image. The source dataset consists of synthetically generated depth images from synthetic RGB images. The target dataset consists of real-world RGB images whose depth images are unknown. An adversarial translation network is trained to convert synthetic RGB images to depth images using the source dataset. An image style transfer network is trained to convert real-world RGB images to synthetic RGB images. The translation and style transfer network are used to estimate the depth version of the real-world RGB images. This chapter is partially based on earlier work from [1].

- **Part IV**: This part discusses novel trends and future directions in domain adaptation research. It consists of 5 chapters with topics including noisy labels, multi-source domains, open set constraints and zero-shot learning.

Chapter 9 discusses a domain adaptation procedure to handle noisy label settings. Their procedure reduces the amount of human supervision required for label noise cleaning. They uniquely pose the problem of domain adaptation as a noisy label problem for the target domain and apply 'attention'-based models to solve large scale problems with label noise. This chapter is partially based on earlier work from [44].

Chapter 10 proposes a novel paradigm for domain adaptation based on Open Set Recognition. In this paradigm the label space of the source is a subset of the label space of the target, which results in the target having known (source) and unknown classes. In order to prevent negative transfer only data points from known classes are aligned from both the domains. The chapter presents a novel Adversarial model where the Generator network is trained to assign a target data point to a known category or declare it as an unknown class. This chapter is partially based on earlier work from [72].

Chapter 11 introduces the Universal Domain Adaptation model which is a generalization of the Open Set domain adaptation paradigm. In standard unsupervised domain adaptation, the labels of the target data are unknown but the label space of the target domain is identical to the label space of the source domain. In the Universal Domain Adaptation setting, the label space of the target is unknown. It could either be a subset, superset or identical to the source label space. The chapter proposes a criterion to determine the transferability of each sample using domain similarity and prediction uncertainty. Using this criterion the model is able to categorize target samples as belonging to the known categories (common with the source) or unknown categories. This chapter is partially based on earlier work from [98].

Chapter 12 proposes a multi-source domain adaptation procedure through a deep CockTail network. Given a target sample, the procedure uses multiple source category classifiers to determine the class assignments. A multiple-source domain discriminator produces perplexity scores that are used to weight the multiple domains in the source. Analogous to making cocktails, the target category is predicted using a mixture of multiple source predictions weighted by their perplexities. The domain discriminator network and the feature-extractor-classifier network are trained in an alternating manner to estimate the target labels. This chapter is partially based on earlier work from [94].

Chapter 13 extends beyond domain adaptation and discusses transfer between tasks. In this chapter the authors introduce a novel zero-shot task transfer model where knowledge from related tasks is used to regress mode parameters for a new task. The model determines pair-wise relations between known tasks and the zero-shot task based on crowd-sourced inputs that are integrated using the Dawid-Skene vote aggregation algorithm. This meta-learning algorithm is able to regress high-performing model parameters for vision tasks without the use of labeled data. This chapter is partially based on earlier work from [59].

References

1. Atapour-Abarghouei, A., Breckon, T.P.: Real-time monocular depth estimation using synthetic data with domain adaptation via image style transfer. In: Proceedings of the IEEE Conference on Computer Vision and Pattern Recognition, pp. 2800–2810 (2018)
2. Baktashmotlagh, M., Faraki, M., Drummond, T., Salzmann, M.: Learning factorized representations for open-set domain adaptation. In: International Conference on Learning Representations (ICLR) (2018)
3. Baxter, J.: A bayesian/information theoretic model of learning to learn via multiple task sampling. Mach. Learn. **28**(1), 7–39 (1997)
4. Bengio, Y.: Learning deep architectures for AI. Found. Trends® Mach. Learn. **2**(1), 1–127 (2009)
5. Bickel, S., Brückner, M., Scheffer, T.: Discriminative learning under covariate shift. J. Mach. Learn. Res. **10**, 2137–2155 (2009)
6. Bruzzone, L., Marconcini, M.: Domain adaptation problems: a DASVM classification technique and a circular validation strategy. IEEE Trans. Pattern Anal. Mach. Intell. **32**(5), 770–787 (2010)
7. Cao, Z., Long, M., Wang, J., Jordan, M.I.: Partial transfer learning with selective adversarial networks. In: Proceedings of the IEEE Conference on Computer Vision and Pattern Recognition (CVPR), pp. 2724–2732 (2018)
8. Cao, Z., Ma, L., Long, M., Wang, J.: Partial adversarial domain adaptation. In: Proceedings of the European Conference on Computer Vision (ECCV), pp. 135–150 (2018)
9. Cao, Z., You, K., Long, M., Wang, J., Yang, Q.: Learning to transfer examples for partial domain adaptation. In: Proceedings of the IEEE Conference on Computer Vision and Pattern Recognition (CVPR), pp. 2985–2994 (2019)
10. Caruana, R.: Multitask learning. Mach. Learn. **28**(1), 41–75 (1997)
11. Chapelle, O., Scholkopf, B., Zien, A. (eds.): Semi-Supervised Learning. MIT Press, Cambridge (2006)
12. Chattopadhyay, R., Sun, Q., Fan, W., Davidson, I., Panchanathan, S., Ye, J.: Multisource domain adaptation and its application to early detection of fatigue. ACM Trans. Knowl. Discov. Data **6**(4), 18 (2012)
13. Chen, Q., Liu, Y., Wang, Z., Wassell, I., Chetty, K.: Re-weighted adversarial adaptation network for unsupervised domain adaptation. In: Proceedings of the IEEE Conference on Computer Vision and Pattern Recognition, pp. 7976–7985 (2018)
14. Daumé III, H., Kumar, A., Saha, A.: Frustratingly easy semi-supervised domain adaptation. In: Workshop on Domain Adaptation for NLP (2010). http://hal3.name/docs/#daume10easyss
15. De Lange, M., Aljundi, R., Masana, M., Parisot, S., Jia, X., Leonardis, A., Slabaugh, G., Tuytelaars, T.: Continual learning: a comparative study on how to defy forgetting in classification tasks (2019). Preprint. arXiv:1909.08383
16. Devlin, J., Chang, M.W., Lee, K., Toutanova, K.: Bert: Pre-training of deep bidirectional transformers for language understanding (2018). Preprint. arXiv:1810.04805
17. Doersch, C., Gupta, A., Efros, A.A.: Unsupervised visual representation learning by context prediction. In: Proceedings of the IEEE International Conference on Computer Vision (ICCV), pp. 1422–1430 (2015)
18. Dudík, M., Phillips, S.J., Schapire, R.E.: Correcting sample selection bias in maximum entropy density estimation. In: Advances in Neural Information Processing Systems (NIPS), pp. 323–330 (2005)
19. Elhoseiny, M., Elfeki, M.: Creativity inspired zero-shot learning. In: Proceedings of the IEEE International Conference on Computer Vision, pp. 5784–5793 (2019)
20. Fei, G., Wang, S., Liu, B.: Learning cumulatively to become more knowledgeable. In: Proceedings of the ACM SIGKDD International Conference on Knowledge Discovery and Data Mining, pp. 1565–1574. ACM, New York (2016)

21. Fei-Fei, L., Fergus, R., Perona, P.: One-shot learning of object categories. IEEE Trans. Pattern Anal. Mach. Intell. **28**(4), 594–611 (2006)
22. Fernando, B., Habrard, A., Sebban, M., Tuytelaars, T.: Unsupervised visual domain adaptation using subspace alignment. In: Proceedings of the IEEE Conference on Computer Vision and Pattern Recognition (CVPR), pp. 2960–2967 (2013)
23. Ganin, Y., Ustinova, E., Ajakan, H., Germain, P., Larochelle, H., Laviolette, F., Marchand, M., Lempitsky, V.: Domain-adversarial training of neural networks. J. Mach. Learn. Res. **17**(59), 1–35 (2016)
24. Geng, C., Huang, S.j., Chen, S.: Recent advances in open set recognition: a survey (2018). Preprint. arXiv:1811.08581
25. Gidaris, S., Singh, P., Komodakis, N.: Unsupervised representation learning by predicting image rotations. In: International Conference on Learning Representations (ICLR) (2018)
26. Gong, B., Shi, Y., Sha, F., Grauman, K.: Geodesic flow kernel for unsupervised domain adaptation. In: Proceedings of the IEEE Conference on Computer Vision and Pattern Recognition (CVPR) (2012)
27. Goodfellow, I., Pouget-Abadie, J., Mirza, M., Xu, B., Warde-Farley, D., Ozair, S., Courville, A., Bengio, Y.: Generative adversarial nets. In: Advances in Neural Information Processing Systems (NIPS), pp. 2672–2680 (2014)
28. Goodfellow, I., Bengio, Y., Courville, A.: Deep learning. MIT Press, Cambridge (2016). http://www.deeplearningbook.org
29. Gopalan, R., Li, R., Chellappa, R.: Domain adaptation for object recognition: an unsupervised approach. In: Proceedings of the IEEE International Conference on Computer Vision (ICCV), pp. 999–1006. IEEE, Piscataway (2011)
30. Grandvalet, Y., Bengio, Y.: Semi-supervised learning by entropy minimization. In: Advances in Neural Information Processing Systems, pp. 529–536 (2005)
31. Gretton, A., Smola, A., Huang, J., Schmittfull, M., Borgwardt, K., Schölkopf, B.: Covariate shift by kernel mean matching. Dataset Shift Mach. Learn. **3**(4), 5 (2009)
32. Heckman, J.J.: Sample selection bias as a specification error. Econometrica **47**(1), 153–161 (1979)
33. Hoffman, J., Kulis, B., Darrell, T., Saenko, K.: Discovering latent domains for multisource domain adaptation. In: Proceedings of the European Conference on Computer Vision (ECCV), pp. 702–715 (2012)
34. Hoffman, J., Rodner, E., Donahue, J., Saenko, K., Darrell, T.: Efficient learning of domain-invariant image representations. In: International Conference on Learning Representations (ICLR) (2013)
35. Hoffman, J., Tzeng, E., Donahue, J., Jia, Y., Saenko, K., Darrell, T.: One-shot adaptation of supervised deep convolutional models (2013). Preprint. arXiv:1312.6204
36. Hu, L., Kan, M., Shan, S., Chen, X.: Duplex generative adversarial network for unsupervised domain adaptation. In: Proceedings of the IEEE Conference on Computer Vision and Pattern Recognition, pp. 1498–1507 (2018)
37. Huang, J., Gretton, A., Borgwardt, K.M., Schölkopf, B., Smola, A.J.: Correcting sample selection bias by unlabeled data. In: Advances in Neural Information Processing Systems (NIPS), pp. 601–608 (2006)
38. Iqbal, J., Ali, M.: Mlsl: Multi-level self-supervised learning for domain adaptation with spatially independent and semantically consistent labeling (2019). Preprint. arXiv:1909.13776
39. Jing, L., Tian, Y.: Self-supervised visual feature learning with deep neural networks: a survey (2019). Preprint. arXiv:1902.06162
40. Joachims, T.: Transductive inference for text classification using support vector machines. In: Proceedings of the ACM International Conference on Machine Learning (ICML), vol. 99, pp. 200–209 (1999)
41. Kingma, D.P., Welling, M.: Auto-encoding variational bayes (2013). Preprint. arXiv:1312.6114
42. Larochelle, H., Erhan, D., Bengio, Y.: Zero-data learning of new tasks. In: Proceedings of the AAAI Conference on Artificial Intelligence, vol. 1, p. 3 (2008)

43. Lee, H., Grosse, R., Ranganath, R., Ng, A.Y.: Convolutional deep belief networks for scalable unsupervised learning of hierarchical representations. In: Proceedings of the ACM International Conference on Machine Learning (ICML), pp. 609–616 (2009)

44. Lee, K.H., He, X., Zhang, L., Yang, L.: Cleannet: transfer learning for scalable image classifier training with label noise. In: Proceedings of the IEEE Conference on Computer Vision and Pattern Recognition, pp. 5447–5456 (2018)

45. Li, Z., Hoiem, D.: Learning without forgetting. In: Proceedings of the European Conf. on Computer Vision (ECCV), pp. 614–629. Springer (2016)

46. Liu, H., Cao, Z., Long, M., Wang, J., Yang, Q.: Separate to adapt: open set domain adaptation via progressive separation. In: Proceedings of the IEEE Conference on Computer Vision and Pattern Recognition (CVPR), pp. 2927–2936 (2019)

47. Long, M., Cao, Y., Wang, J., Jordan, M.: Learning transferable features with deep adaptation networks. In: Proceedings of the ACM International Conference on Machine Learning (ICML), pp. 97–105 (2015)

48. Long, M., Wang, J., Ding, G., Sun, J., Yu, P.S.: Transfer feature learning with joint distribution adaptation. In: Proceedings of the ACM International Conference on Machine Learning (ICML), pp. 2200–2207 (2013)

49. Long, M., Wang, J., Ding, G., Sun, J., Yu, P.S.: Transfer joint matching for unsupervised domain adaptation. In: Proceedings of the IEEE Conference on Computer Vision and Pattern Recognition (CVPR) (2014)

50. Mairal, J., Bach, F., Ponce, J., Sapiro, G.: Online learning for matrix factorization and sparse coding. J. Mach. Learn. Res. **11**, 19–60 (2010)

51. Mansour, Y., Mohri, M., Rostamizadeh, A.: Domain adaptation with multiple sources. In: Advances in Neural Information Processing Systems (NIPS), pp. 1041–1048 (2009)

52. Mensink, T., Verbeek, J., Perronnin, F., Csurka, G.: Distance-based image classification: generalizing to new classes at near-zero cost. IEEE Trans. Pattern Anal. Mach. Intell. **35**(11), 2624–2637 (2013)

53. Misra, I., Shrivastava, A., Gupta, A., Hebert, M.: Cross-stitch networks for multi-task learning. In: Proceedings of the IEEE Conference on Computer Vision and Pattern Recognition, pp. 3994–4003 (2016)

54. Misra, I., Zitnick, C.L., Hebert, M.: Shuffle and learn: unsupervised learning using temporal order verification. In: Proceedings of the European Conference on Computer Vision (ECCV), pp. 527–544. Springer, Berlin (2016)

55. Motiian, S., Jones, Q., Iranmanesh, S., Doretto, G.: Few-shot adversarial domain adaptation. In: Advances in Neural Information Processing Systems, pp. 6670–6680 (2017)

56. Murez, Z., Kolouri, S., Kriegman, D., Ramamoorthi, R., Kim, K.: Image to image translation for domain adaptation. In: Proceedings of the IEEE Conference on Computer Vision and Pattern Recognition, pp. 4500–4509 (2018)

57. Ng, A.: Hiring your first chief AI officer. Harvard Business Review (2016). https://hbr.org/2016/11/hiring-your-first-chief-ai-officer

58. Nowozin, S., Lampert, C.H.: Structured learning and prediction in computer vision. Found. Trends Comput. Graph. Vis. **6**(3–4), 185–365 (2011). https://doi.org/10.1561/0600000033

59. Pal, A., Balasubramanian, V.N.: Zero-shot task transfer. In: Proceedings of the IEEE Conference on Computer Vision and Pattern Recognition, pp. 2189–2198 (2019)

60. Palatucci, M., Pomerleau, D., Hinton, G.E., Mitchell, T.M.: Zero-shot learning with semantic output codes. In: Advances in Neural Information Processing Systems (NIPS), pp. 1410–1418 (2009)

61. Pan, S.J., Yang, Q.: A survey on transfer learning. IEEE Trans. Knowl. Data Eng. **22**(10), 1345–1359 (2010)

62. Pan, S.J., Kwok, J.T., Yang, Q.: Transfer learning via dimensionality reduction. In: Proceedings of the AAAI Conference on Artificial Intelligence, vol. 8, pp. 677–682 (2008)

63. Pan, S.J., Tsang, I.W., Kwok, J.T., Yang, Q.: Domain adaptation via transfer component analysis. IEEE Trans. Neural Netw. **22**(2), 199–210 (2011)

64. Panareda Busto, P., Gall, J.: Open set domain adaptation. In: Proceedings of the IEEE International Conference on Computer Vision (ICCV), pp. 754–763 (2017)
65. Parisi, G.I., Kemker, R., Part, J.L., Kanan, C., Wermter, S.: Continual lifelong learning with neural networks: a review. Neural Netw. **113**, 54–71 (2019)
66. Quionero-Candela, J., Sugiyama, M., Schwaighofer, A., Lawrence, N.D.: Dataset shift in machine learning. The MIT Press, Cambridge (2009)
67. Raina, R., Battle, A., Lee, H., Packer, B., Ng, A.Y.: Self-taught learning: transfer learning from unlabeled data. In: Proceedings of the ACM International Conference on Machine Learning (ICML), pp. 759–766 (2007)
68. Rebuffi, S.A., Kolesnikov, A., Lampert, C.H.: iCaRL: incremental classifier and representation learning. In: Proceedings of the IEEE Conference on Computer Vision and Pattern Recognition (CVPR) (2017)
69. Ruder, S.: An overview of multi-task learning in deep neural networks (2017). Preprint. arXiv:1706.05098
70. Ruder12, S., Bingel, J., Augenstein, I., Søgaard, A.: Sluice networks: learning what to share between loosely related tasks. Comp. Sci. Math. **1050**, 23 (2017)
71. Saenko, K., Kulis, B., Fritz, M., Darrell, T.: Adapting visual category models to new domains. In: Proceedings of the European Conf. on Computer Vision (ECCV) (2010)
72. Saito, K., Yamamoto, S., Ushiku, Y., Harada, T.: Open set domain adaptation by backpropagation. In: Proceedings of the European Conference on Computer Vision (ECCV), pp. 153–168 (2018)
73. Schoenauer-Sebag, A., Heinrich, L., Schoenauer, M., Sebag, M., Wu, L.F., Altschuler, S.J.: Multi-domain adversarial learning (2019). Preprint. arXiv:1903.09239
74. Shimodaira, H.: Improving predictive inference under covariate shift by weighting the log-likelihood function. J. Stat. Plann. Inference **90**(2), 227–244 (2000)
75. Socher, R., Ganjoo, M., Manning, C.D., Ng, A.: Zero-shot learning through cross-modal transfer. In: Advances in Neural Information Processing Systems (NIPS), pp. 935–943 (2013)
76. Sugiyama, M., Nakajima, S., Kashima, H., Buenau, P.V., Kawanabe, M.: Direct importance estimation with model selection and its application to covariate shift adaptation. In: Advances in Neural Information Processing Systems (NIPS), pp. 1433–1440 (2008)
77. Sukhija, S., Krishnan, N.C.: Supervised heterogeneous feature transfer via random forests. Artif. Intell. **268**, 30–53 (2019)
78. Sukhija, S., Krishnan, N.C., Kumar, D.: Supervised heterogeneous transfer learning using random forests. In: Proceedings of the ACM India Joint International Conference on Data Science and Management of Data, pp. 157–166 (2018)
79. Sun, Q., Chattopadhyay, R., Panchanathan, S., Ye, J.: A two-stage weighting framework for multi-source domain adaptation. In: Advances in Neural Information Processing Systems (NIPS), pp. 505–513 (2011)
80. Tan, S., Jiao, J., Zheng, W.S.: Weakly supervised open-set domain adaptation by dual-domain collaboration. In: Proceedings of the IEEE Conf. on Computer Vision and Pattern Recognition (CVPR), pp. 5394–5403 (2019)
81. Tarvainen, A., Valpola, H.: Mean teachers are better role models: weight-averaged consistency targets improve semi-supervised deep learning results. In: Advances in Neural Information Processing Systems, pp. 1195–1204 (2017)
82. Thrun, S.: Is learning the n-th thing any easier than learning the first? In: Advances in Neural Information Processing Systems, pp. 640–646. Morgan Kaufmann Publishers, Burlington (1996)
83. Thrun, S., Pratt, L.: Learning to Learn. Springer Science & Business Media, Berlin (2012)
84. Tzeng, E., Hoffman, J., Zhang, N., Saenko, K., Darrell, T.: Deep domain confusion: maximizing for domain invariance (2014). Preprint. arXiv:1412.3474
85. Vaezi Joze, H.R., Shaban, A., Iuzzolino, M.L., Koishida, K.: MMTM: multimodal transfer module for CNN fusion. In: Proceedings of the IEEE Conference on Computer Vision and Pattern Recognition (2020)

86. van den Oord, A., Kalchbrenner, N., Kavukcuoglu, K.: Pixel recurrent neural networks (2016). Preprint. arXiv:1601.06759
87. Venkatesan, R., Venkateswara, H., Panchanathan, S., Li, B.: A strategy for an uncompromising incremental learner (2017). Preprint. arXiv:1705.00744
88. Venkateswara, H., Lade, P., Ye, J., Panchanathan, S.: Coupled support vector machines for supervised domain adaptation. In: Proceedings of the ACM International Conference on Multimedia (ACM-MM), pp. 1295–1298 (2015)
89. Venkateswara, H., Chakraborty, S., Panchanathan, S.: Nonlinear embedding transform for unsupervised domain adaptation. In: Workshops, Proceedings of the European Conf. on Computer Vision (ECCV TASK-CV), pp. 451–457. Springer, Berlin (2016)
90. Venkateswara, H., Chakraborty, S., Panchanathan, S.: Deep-learning systems for domain adaptation in computer vision: learning transferable feature representations. IEEE Signal Process. Mag. **34**(6), 117–129 (2017)
91. Venkateswara, H., Eusebio, J., Chakraborty, S., Panchanathan, S.: Deep hashing network for unsupervised domain adaptation. In: Proceedings of the IEEE Conference on Computer Vision and Pattern Recognition (CVPR) (2017)
92. Wang, J., Jiang, J.: Conditional coupled generative adversarial networks for zero-shot domain adaptation. In: Proceedings of the IEEE International Conference on Computer Vision (ICCV), pp. 3375–3384 (2019)
93. Woodworth, R.S., Thorndike, E.: The influence of improvement in one mental function upon the efficiency of other functions (i). Psychoanal. Rev. **8**(3), 247 (1901)
94. Xu, R., Chen, Z., Zuo, W., Yan, J., Lin, L.: Deep cocktail network: multi-source unsupervised domain adaptation with category shift. In: Proceedings of the IEEE Conference on Computer Vision and Pattern Recognition, pp. 3964–3973 (2018)
95. Xu, X., Zhou, X., Venkatesan, R., Swaminathan, G., Majumder, O.: d-SNE: Domain adaptation using stochastic neighborhood embedding. In: Proceedings of the IEEE Conference on Computer Vision and Pattern Recognition, pp. 2497–2506 (2019)
96. Yang, Y., Hospedales, T.M.: Trace norm regularised deep multi-task learning (2016). Preprint. arXiv:1606.04038
97. Yang, J., Yu, K., Gong, Y., Huang, T.: Linear spatial pyramid matching using sparse coding for image classification. In: Proceedings of the IEEE Conference on Computer Vision and Pattern Recognition (CVPR), pp. 1794–1801 (2009)
98. You, K., Long, M., Cao, Z., Wang, J., Jordan, M.I.: Universal domain adaptation. In: Proceedings of the IEEE Conference on Computer Vision and Pattern Recognition (CVPR), pp. 2720–2729 (2019)
99. Zadrozny, B.: Learning and evaluating classifiers under sample selection bias. In: Proceedings of the ACM International Conference on Machine Learning (ICML), p. 114 (2004)
100. Zhang, R., Isola, P., Efros, A.A.: Colorful image colorization. In: Proceedings of the European Conference on Computer Vision (ECCV), pp. 649–666. Springer, Berlin (2016)
101. Zhang, J., Ding, Z., Li, W., Ogunbona, P.: Importance weighted adversarial nets for partial domain adaptation. In: Proceedings of the IEEE Conference on Computer Vision and Pattern Recognition (CVPR), pp. 8156–8164 (2018)
102. Zhao, S., Wang, G., Zhang, S., Gu, Y., Li, Y., Song, Z., Xu, P., Hu, R., Chai, H., Keutzer, K.: Multi-source distilling domain adaptation. In: Proceedings of the AAAI Conference on Artificial Intelligence (2020)
103. Zhu, Y., Elhoseiny, M., Liu, B., Peng, X., Elgammal, A.: A generative adversarial approach for zero-shot learning from noisy texts. In: Proceedings of the IEEE Conference on Computer Vision and Pattern Recognition, pp. 1004–1013 (2018)
104. Žliobaitė, I., Pechenizkiy, M., Gama, J.: An overview of concept drift applications. In: Big Data Analysis: New Algorithms for a New Society, pp. 91–114. Springer, Berlin (2016)

Chapter 2
Shallow Domain Adaptation

Sanatan Sukhija and Narayanan Chatapuram Krishnan

2.1 Introduction

Traditional supervised algorithms learn a predictor from the available labeled examples with the goal of generalizing well to unseen examples. Learning a robust predictor requires a large number of labeled examples. Often, in several real-world scenarios, the quantity of available labeled examples is scarce. Moreover, even if unlabeled examples are present, manually labeling them demands domain expertise and is a laborious task. Leveraging labeled or unlabeled knowledge from existing auxiliary domains (often termed as the source domains) can mitigate these shortcomings to learn the desired task efficiently in the target domain. Knowledge transfer between two domains is primarily challenged by the differences in the underlying data distributions, the feature spaces and the output spaces.

Figure 2.1 presents a generic categorization of the knowledge transfer settings based on the characteristics of the source and target domain. When the source and target data belong to the same feature space having the same underlying distribution, the transfer scenario becomes equivalent to the traditional supervised learning setting. In contrast, when the target distribution differs, the distributions have to be matched so that the source model can perform well on the target examples. The transfer setting is more commonly referred to as Domain Adaptation (DA). Mainly, the objective of the DA approaches is to minimize the marginal distribution differences between the domains. Given a few labeled examples in the target domain, the conditional distributions can also be matched. The knowledge transfer task for domains having heterogeneous features is more commonly known as Heterogeneous Transfer Learning or Heterogeneous Domain Adaptation (HDA).

S. Sukhija · N. Chatapuram Krishnan (✉)
Indian Institute of Technology Ropar, Rupnagar, Punjab, India
e-mail: sanatan@iitrpr.ac.in; ckn@iitrpr.ac.in

© Springer Nature Switzerland AG 2020
H. Venkateswara, S. Panchanathan (eds.), *Domain Adaptation in Computer Vision with Deep Learning*, https://doi.org/10.1007/978-3-030-45529-3_2

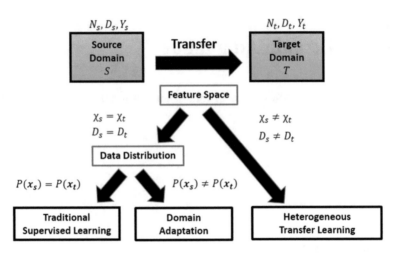

Fig. 2.1 Different machine learning paradigms

This chapter summarizes the state-of-the-art generic shallow (non-deep learning) DA and HDA approaches from the pre-deep learning era.

Consider a real-world sentiment classification task with the goal of identifying the sentiment of electronic product reviews that are written in the Spanish language. A review expresses either a positive sentiment or a negative sentiment towards the product experience. Learning a robust predictive function requires abundant Spanish reviews specific to electronic products. Another indispensable requirement is the preliminary processing of the collected reviews such as labeling the sentiment of the reviews, removing fuzzy reviews from the dataset, defining the input vocabulary (features), etc., which might require an expert(s) proficient in Spanish. Moreover, if there are not enough labeled reviews available for training, the learned function cannot reliably predict the sentiment. The task can benefit from leveraging labeled or unlabeled reviews from related auxiliary sources. The labeled or unlabeled Spanish reviews of other products such as books, music, etc. or electronic product reviews in other languages such as English that are easily available can assist in the task. Here, utilizing labeled Spanish reviews of other products such as books, music, etc., corresponds to the first transfer scenario of DA, whereas the second scenario of leveraging labeled English reviews corresponds to the HDA setting.

2.1.1 Chapter Organization

After the informal introduction to knowledge transfer with the help of a motivating application in Sects. 2.1 and 2.1.2 describes the notations of the terminologies used in the adaptation literature. The next Sect. 2.1.3, gives the formal definition of the different adaptation scenarios. Section 2.1.4 reports a few real-world applications that are well-known in the shallow adaptation literature for benchmarking the performance of adaptation algorithms. Sections 2.2 and 2.3 presents a compact review of the promising DA and HDA approaches respectively. The last section (Sect. 2.4) highlights the unexplored avenues in the adaptation research. This chapter is partially based on earlier work from [60, 62].

2.1.2 Notations

A domain represents a set of examples $\{(\mathbf{x}_1, y_1), \ldots, (\mathbf{x}_N, y_N)\}$ that are sampled from the marginal probability distribution $P(\mathbf{x})$. Here, each example $\mathbf{x}_i$ belongs to the feature space $\mathcal{X} \in \mathbb{R}^D$ with $y_i \in Y$ as its label. A task T is described as learning a predictor $f(.)$ from the available labeled examples that maps the features $\mathcal{X}$ to the labels Y. Given an unseen example $\mathbf{x}$, the learned function $f(.)$ predicts the label $f(\mathbf{x}) \in Y$. From the Bayesian perspective, the predictor $f(.)$ can be perceived as a function that outputs the probability $P(y|\mathbf{x})$.

For simplicity, we describe the notations for the learning setting when there is a single source domain. The source domain S contains N_s examples $\{(\mathbf{x}_i^s, y_i^s)\}_{i=1}^{N_s}$ that are sampled from the marginal distribution $P(\mathbf{x}_s)$ and the conditional distribution $P(y_s|\mathbf{x}_s)$. Each labeled example $\mathbf{x}_i^s$ in the source feature space $\mathcal{X}_s$ is represented with a D_s-dimensional vector having $y_i^s \in Y_s$ as its corresponding label. In the target domain T, we have N_t examples $\{(\mathbf{x}_j^t, y_j^t)\}_{j=1}^{N_t}$ having the marginal distribution $P(\mathbf{x}_t)$ and the conditional distribution $P(y_t|\mathbf{x}_t)$. A target example $\mathbf{x}_j^t \in \mathcal{X}_t$ is represented with a D_t-dimensional vector with $y_j^t \in Y_t$ as its corresponding label. T_s and T_t denote the source and target task respectively.

2.1.3 Formal Definitions

In general, the source domain contains a large number of labeled examples. However, depending on whether labeled examples are available in the target domain, the transfer learning frameworks are associated with either of the three configurations:

- **Unsupervised**: No labeled examples in the target domain.
- **Supervised**: The approaches in this category utilize only the labeled examples from the target domain.
- **Semi-Supervised**: Both labeled and unlabeled examples from the target domain are utilized for adaptation.

According to Pan et al.'s description of the transfer learning framework [46], there are two possible transfer scenarios, where the domains or the tasks are different. The first scenario with $S \neq T$ means that either $\mathcal{X}_s \neq \mathcal{X}_t$ or $P_s(\mathbf{x}_s) \neq$

$P_t(\mathbf{x}_t)$ and the second scenario with $T_s \neq T_t$ indicates that either $Y_s \neq Y_t$ or $P(y_s|\mathbf{x}_s) \neq P(y_t|\mathbf{x}_t)$. Even if we say that the tasks or domains are different, most conventional transfer learning methods assume that they are related to a certain extent so as to avoid negative transfer [45].

Definition 2.1 (Domain Adaptation) The objective here is to learn a reliable predictor $f_t(.)$ for T by leveraging the labeled examples from S or the knowledge from T_s. Under this setting, the source and target domain share the same feature space $\mathcal{X}_s = \mathcal{X}_t$ and also, the same task $T_s = T_t$. However, the domains are different $S \neq T$ in terms of the marginal distributions $P(\mathbf{x}_s) \neq P(\mathbf{x}_t)$.

Definition 2.2 (Heterogeneous Domain Adaptation) With the same goal as DA, the only difference is that the domains differ $S \neq T$ in terms of the feature spaces $\mathcal{X}_s \neq \mathcal{X}_t$.

2.1.4 Applications

In this section, we have described some real-world adaptation tasks that are prominent in the pre-deep learning literature.

- **Visual Domain Adaptation**: The task of object recognition deals with identifying objects in a given image. The differences across visual domains such as lighting conditions, color, resolution, background, camera orientation, etc. result in distribution mismatch, which calls for adaptation. In the shallow visual adaptation literature, knowledge transfer has been performed within and across different kinds of Image descriptors such as Speeded Up Robust Features, Classemes and Scale-Invariant Feature Transform, etc. Knowledge transfer across different image datasets for object recognition task is popularly known as cross-domain object recognition.
- **Cross Domain Text/Sentiment Classification**: As mentioned in the motivating application earlier, relevant labeled documents from other domains can be utilized to learn the target task. The use of labeled documents from homogeneous auxiliary domains causes domain induced bias due to word-distribution differences whereas utilizing labeled documents from other languages necessitates handling vocabulary differences first. As knowledge transfer occurs across different languages, the latter setting is more commonly known as Cross-lingual text/sentiment classification.
- **Cross-modal classification task**: Given the same task in both domains, knowledge transfer across domains having different modalities such as Image $\leftrightarrow$ Text is known as a cross-modal classification task. This task is closely related to the Multi-View learning [39] setting where complementary information from other modalities (views) is simultaneously available in the form of parallel data to learn more comprehensive representations for the desired task.

- **Cross Domain Activity Recognition**: The task of identifying the actions per-formed by the resident(s) of a smart home from the recorded sensor observations is known as smart home activity recognition. Training a robust model demands manually labeling the sensor records, which is time-consuming and laborious. Transfer learning approaches can be used to leverage labeled observations from existing smart homes to a target smart home to circumvent the annotation effort. The usage of different types of sensors across the different smart home layouts leads to heterogeneous feature spaces. The knowledge transfer task across different smart homes is commonly known as Cross-Domain Activity Recognition.
- **Cross-Domain Indoor WiFi Localization**: The task of accurately tracking and locating a mobile object with WiFi signal strength has garnered a lot of attention lately [83]. The objective is to determine the whereabouts of a mobile appliance with the help of signal strengths received from different access nodes. In contrast to previous applications, the indoor WiFi localization task is a regression problem. This task requires timely calibration as the received signal strengths vary with traffic and time. In order to avoid the expensive calibration task for a large environment, domain adaptation is employed.

2.2 Domain Adaptation

The performance of a supervised model is validated on test examples that are drawn from the same distribution as that of training examples. However, when the learned model is evaluated on examples from a different domain, the performance suffers due to the distribution differences between the domain data [66]. Shai et al. [7] stated that the performance of a classifier in a new domain (the target domain) depends on two factors: (1) the performance in its own domain (the source domain) and (2) the discrepancy between the domains. Given identical labels $Y_s = Y_t$ and the same feature space $\mathcal{X}_s = \mathcal{X}_t$, Domain Adaptation (DA) deals with minimizing the discrepancy between the domains. The domain discrepancy can be in terms of the differences in the marginal distributions $P_s(\mathbf{x}_s) \neq P_t(\mathbf{x}_t)$ or the conditional distributions $P(y_s|\mathbf{x}_s) \neq P(y_t|\mathbf{x}_t)$.

Given very few labeled examples in the target domain (or no labeled examples at all), it is hard to estimate $P_t(y_t|\mathbf{x}_t)$. Most DA approaches assume the domains to be correlated to some extent with $P(y_s|\mathbf{x}_s) \approx P(y_t|\mathbf{x}_t)$ and only focus on minimizing the marginal distribution differences. This setting is known with many names in the adaptation literature, namely, Domain Shift, Dataset Shift, and Covariate Shift. However, when the conditional distribution differences are also significant $P(y_s|\mathbf{x}_s) \neq P(y_t|\mathbf{x}_t)$, then this setting is more commonly known as Sample Selection Bias. We summarize the shallow DA approaches in the next section.

2.2.1   A Condensed Review of DA Literature

The Support Vector Machine (SVM) [70] based frameworks are semi-supervised
DA approaches that modify the SVM optimization framework to use labeled
examples from both the domains simultaneously for adapting to the target domain
data. Yang et al. [81] proposed Adaptive SVM where the SVM source model is
regularized for the target examples. Domain Adaptive Support Vector Machine [12]
and Projective Model Transfer SVM [3] are very similar adaptation approaches
where the source SVM decision boundary is iteratively adapted to better classify
the target examples. Wu et al. [79] proposed another SVM-based framework to
determine relevant support vectors from the source domain for the target task.
Hoffman et al. introduced Maximum Margin Domain Transfer [31] where a linear
transformation that maps the target features to the source features is learned to
account for the distribution mismatch.

   The Subspace-based adaptation methods are Unsupervised DA methods that
align the source and target features. The Subspace Alignment method [20] has a
closed-form solution that aligns the Eigen feature spaces obtained through PCA.
Similarly, manifold based DA methods [25, 27] minimize the distribution differences in a low-dimensional manifold. In contrast to the aforementioned subspace-
based methods that select the foremost subspace dimensions for alignment, the
Correlation Alignment [64] approach directly minimizes the covariances of the
original feature space. Another approach by Si et al. [57] minimizes the Bregman
divergence between the distributions in the common subspace. In contrast to the
above approaches, Daume et al. [15] proposed Easy Adapt(EA), a semi-supervised
DA approach that uses a feature augmentation strategy for the domains to define
a common subspace which comprises of three components: domain-independent,
source-specific and target-specific features. The extension of EA, EA++ [16] uses
unlabeled examples in the target domain for learning better hypotheses for the target
domain. This is achieved by forcing the source and target hypothesis to agree on the
predictions made on the unlabeled target examples.

   The instance-transfer based sample selection strategies are unsupervised iterative
learning frameworks that learn the importance (weight) of the samples towards
the target task. In the covariate shift setting, the weight of a sample $w(x)$ should
be set to the ratio between the marginal probabilities of the target and the source
domain i.e. $\frac{P_t(x)}{P_s(x)}$ [68]. Early instance re-weighting strategies estimated this ratio
by training a domain classifier [86] or with Kullback-Leibler divergence [36, 58]
and Least-squared loss [37] as the loss function. Another popular weighting scheme
[29, 33] is to use marginal distribution difference between the domains computed
with the Maximum Mean Discrepancy (MMD) [10] principle. Inspired by boosting
[23], the Transfer Adaptive Boosting (TrAdaBoost) [14] approach increases the
weights of incorrectly classified target samples whereas the weights associated
with correctly classified source samples are decreased while iteratively training the
target classifier. Al-Stouhi and Reddy [2] introduced dynamic weight updates in the
TrAdaBoost to counter premature convergence of source weights. Recent strategies

are formulated as optimization frameworks to identify relevant source examples (also known as landmarks) that are distributed similarly to the target domain. Gong et al. [26] proposed an unsupervised MMD based optimization framework to learn these landmarks. Instead of using MMD, Baktashmotlagh et al. [5] learn these landmarks with the Hellinger distance to minimize the weighted-source and target distributions on a Riemannian manifold (subspace). In contrast to the above unsupervised frameworks, Multi-scale Landmarks Selection (MLS) [1] does not utilize the source labels and the landmarks are computed from both the source and target samples. MLS first identifies the landmarks in the kernel space and then uses these landmarks to perform the alignment in the PCA subspace.

The Feature Transformation approaches minimize the distribution differences in a latent space by learning a common transformation or domain-specific transformation(s). Transfer Component Analysis (TCA) [48] is an unsupervised FT method that uses the MMD principle to minimize the distribution differences between the domains. The Domain Invariant Projection [4] also uses the MMD principle to bridge the distribution differences. However, the algorithm learns an orthogonal transformation to first project the data from both the domains to a common latent space and then aligns the distribution using MMD. Transfer Joint Matching [42] combines TCA with a regularizer term that facilitates adaptive instance re-weighting. The semi-supervised version of TCA [43] introduces a label dependency term in the TCA framework that utilizes the source and target labels to induce class-wise alignment. The Adaptation Regularization based Transfer Learning framework [41] is a semi-supervised optimization framework that combines (1) structural risk minimization for learning the target classifier from labeled source examples, (2) the MMD principle for minimizing the marginal and (class-wise) conditional distribution differences while retaining (3) manifold consistency [6] for individual domains. Zhong et al. [89] proposed a semi-supervised FT framework that minimizes the marginal distribution differences between the domains in the kernel space and additionally, utilizes a clustering-based strategy to identify relevant source examples for the target task, which in turn helps to minimize the condition distribution differences. Joint Distribution Adaptation [40] learns a transformation to minimize the marginal distribution differences between the source labeled data and target unlabeled data in the PCA subspace. Furthermore, the algorithm utilizes the pseudo-labels (predicted labels) of the target samples that are obtained from a classifier trained on source samples, to reduce the conditional distribution differences in that PCA subspace.

Most of the above DA methods consider a single source domain. Alternatively, when multiple source domains are given, the adapted DA models for every source-target pair can be jointly used to improve the generalization in the target domain. In contrast, learning a Multi-Source DA model in unison is comparatively better as it can simultaneously exploit the domain-specific information from multiple domains. Moreover, it is not necessary that all the source domains are equally relevant for the target task. One simple approach is to use A-distance metric [7] to evaluate relatedness of a source domain with the target domain and then accordingly weight the predictions. Besides EA and EA++ [15, 16] that can be naturally extended to accommodate multiple source domains by augmenting the feature space, other approaches such as Adaptive-SVM [82] trains an ensemble of weighted-source classifiers to adapt for the target task. The Multi-Model Knowledge Transfer framework [65] jointly trains multiple linear SVMs from source domain data in a weighted fashion where the weights are determined by using the leave-one-out method. The Domain Adaptation Machine [19] introduced a new domain-dependent regularization term that ensures that the target classifier (the one trained with labeled target examples) shares similar predictions with pre-trained relevant base classifiers on unlabeled examples in the target domain. Here, the base classifiers are trained either with source domain data or by combining source and target labeled data. Some of the previously described approaches for DA have been extended to the multi-source setting. For e.g. Gopalan et al. [28] extended the manifold based DA method [27] and Yao et al. [85] extended TrAdaboost [14] for the multi-source setting.

2.3 Heterogeneous Domain Adaptation

In practice, it is challenging to find existing data that is represented by the same set of features with a shared label space as that of the target domain. HDA deals with knowledge transfer between domains having heterogeneous features.

2.3.1 A Concise Review of Recent HDA Literature

As the source and target domain feature spaces are heterogeneous in this transfer setting, a preliminary task for every HDA algorithm is to decide a feature space for adaptation. Based on the feature space where adaptation was performed, the HDA approaches can be summarized under two broad categories:

1. Feature Space Remapping (FSR)
2. Latent Space Transformation (LST)

The FSR approaches perform adaptation either in the source feature space or the target feature space. This requires learning a single mapping between the heterogeneous features of the domains i.e. $P_s : \mathcal{X}_s \rightarrow \mathcal{X}_t$ or $P_t : \mathcal{X}_t \rightarrow \mathcal{X}_s$. The learned mapping is then used to project the data from one domain to the other. Figure 2.2 depicts the framework for FSR based TL approaches.

In contrast, the LST approaches project the data from the source and target domain to a shared subspace for adaptation which requires learning two mappings, one for each domain. Figure 2.3 depicts the framework for LST based TL approaches.

Adapting to the target task in the desired feature space needs some shared pivotal information to bridge the differences that arise when the data from the domains is

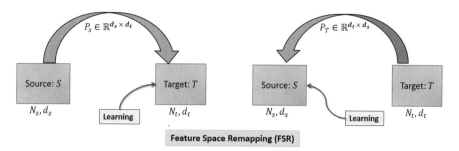

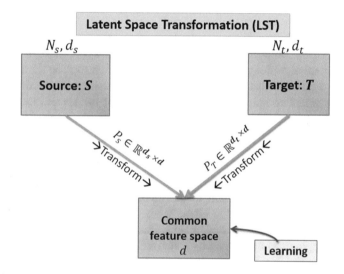

Fig. 2.2 The framework for Feature Space Remapping approaches

Fig. 2.3 The framework for Latent Space Transformation approaches

jointly aligned. Recent HDA approaches have utilized few labeled target examples [34, 59, 61, 62, 76, 78, 80, 90] to match the conditional distributions. Blitzer et al. [8] utilized the overlapping features (common features across the domains) to bridge the remaining heterogeneous features. Some HDA approaches explicitly define a meta-feature space [18] across the domains which can be treated as the common space. Other task-specific HDA approaches [30, 32, 91] jointly partition the heterogeneous feature spaces into domain-independent features and domain-specific features. Some HDA approaches utilize social media/web [77, 78, 84, 92] and oracles/dictionary [24, 51] to leverage domain-specific knowledge in the form of corresponding instances or features to bridge the domain differences. Typically, learning the mapping(s) for HDA is formulated as a minimization objective. Equation 2.1 depicts a generic HDA optimization framework.

$$\min_{P_s, P_t, B_s, B_t, M} L(S, T, B_s, B_t, P_s, P_t, M) \; + \; J(S, T, B_s, B_t, P_s, P_t) \; + $$

$$+ \; R(P_s, P_t, B_s, B_t, M) \; \text{w.r.t} \; C(B_s, B_t, P_s, P_t) \tag{2.1}$$

The minimization framework comprises of four components. The first component $L(.)$ represents the distortion function that computes the loss incurred while trying to project the source data $S \in \mathbb{R}^{N_s \times D_s}$ and the target data $T \in \mathbb{R}^{N_t \times D_t}$ by using the mappings P_s and P_t respectively. Here, B_s and B_t represent the projected source and target domain data. For LST optimization approaches, we have to learn two transformations $P_s \in \mathbb{R}^{D_s \times D}$ and $P_t \in \mathbb{R}^{D_t \times D}$ where D represents the number of dimensions in the shared subspace, whereas for FSR approaches, $P_s \in \mathbb{R}^{D_s \times D_t}$ and $P_t = P_s'$. The second component $J(.)$ is responsible for maximizing the similarity of the jointly aligned data by leveraging some shared information from the domains. Depending on the available shared information, one can compensate for the distribution differences or the topological differences. The third component $R(.)$ is the regularizer that regulates the impact of variables on the final solution, controls the model complexity M, decides sparsity of the learned transformation and avoids over-fitting to limited labeled data. The fourth component $C(.)$ represents the constraints that cater to specific needs of the problem such as orthogonality of projected data or transformation(s), scale-invariance or symmetricity of the learned transformation, etc.

The shallow FSR approaches learn a single linear mapping that directly transforms the data from one domain to the other. Assuming if the objective is to transform the source data to the target space, then the mapping $P_s \in \mathbb{R}^{D_s \times D_t}$ is learned. Given sufficient cross-domain correspondences (also known as parallel data), the mapping can be learned by using multi-output regression [9]. Given the source data matrix $S \in \mathbb{R}^{N \times D_s}$ and the target data matrix $T \in \mathbb{R}^{N \times D_t}$ where N denotes the count of available correspondences, Eq. 2.2 depicts the minimization objective that learns the optimal mapping $P_s \in \mathbb{R}^{D_s \times D_t}$.

$$\min_{P_s} \frac{1}{N} \; || \; T - S P_s \; ||_2^2 + \lambda \; || \; P_s \; ||_2^2 \tag{2.2}$$

Here, λ denotes the regularizer. However, it is relatively difficult to obtain cross-domain correspondences for real-world transfer tasks. The shallow FSR approaches can be further categorized into two scenarios, namely, (1) when there is a one to one feature correspondence across the domains or (2) a feature from the source domain is represented as a linear combination of the features in the target domain (or vice-versa). For heterogeneous transfer tasks, especially involving transfer across different modalities (e.g. text → Image), the feature correspondences between the source and target domain are not explicit. Moreover, even for heterogeneous transfer tasks where the source and target domain have the same modality such as cross-lingual sentiment/text classification [49] (modality: text) or cross-domain activity recognition [61] (modality: sensors), it is a challenging task to determine one-to-one feature correspondences. Feuz et al. [22] proposed two computation intensive strategies, namely, (1) greedily mapping the features across the domains based on some fitness metric and (2) leveraging genetic algorithms to learn the feature correspondences. Given the heterogeneous domains share an explicit subset of features, some task-specific approaches [8, 30, 45, 51] leverage the common features to determine the mapping between the remaining features across the domains. The applicability of these algorithms is limited as it is difficult to find common features across heterogeneous domains for most transfer tasks.

Recent state-of-the-art FSR approaches [34, 60–62, 90] use some labeled target examples to learn the mapping. Cross-Domain Landmark Selection [34] can be summarized as an instance-based sample selection FSR approach for heterogeneous transfer tasks. The approach learns the importance of the source domain examples while matching both the marginal and the conditional distributions of the domains. Other supervised FSR approaches [60–62, 90] generate cross-domain correspondences by leveraging the common labels. Sparse Heterogeneous Feature Remapping (SHFR) [90] approach trains SVM models for both the domains. The corresponding source and target SVM weight vector (belonging to the same class) is considered as a cross-domain corresponding pair. Learning a robust mapping demands plenty of correspondences. SHFR encodes the labels with error-correcting output codes (ECOC) to generate more correspondences. In contrast to using the individual labels, Sukhija et al. [60–62] use label distributions to generate correspondences. The key idea of the approach is that it considers similarly labeled partitions across the domains for generating cross-domain correspondences. The approach trains Random Forest [11] models for the source and target domain to determine these labeled partitions. A tree path contains a sequence of features that were designated to recursively split the data where the leaf node holds a distribution of the class labels. The contribution of the source features towards a source data partition is deemed as the weight vector for that partition. Similarly, for a target data partition, the weight vector is computed using the contribution of only those target features that are used to generate that partition. The corresponding weight vectors for the similarly labeled partitions across the domains are considered as a cross-domain corresponding pair. Similar to SHFR, the algorithm estimates a linear and sparse mapping from the generated correspondences. The aforementioned supervised FSR approaches utilize the common labels to estimate cross-domain correspondences and therefore, these approaches are not directly applicable for knowledge transfer between domains having heterogeneous labels. A recent work by Sukhija et al. [59] bridges heterogeneous domains with semantically related label spaces, where the label relationships are estimated through a web-based similarity metric known as Normalized Google Distance. Based on the contrastive loss, the supervised FSR approach transforms the source data to the target feature space. In the target feature space, semantically similar examples are kept closer whereas the dissimilar ones are kept at a distance while ensuring the original structure of the source data is preserved. The approach also incorporates a sample selection strategy to identify a subset of the transformed source examples that are very similar to the target examples. The obtained source subset is used with the labeled examples from the target domain to train the model for the target task.

The LST approaches bridge the feature divergence gap between the transformed data of the domains in the shared subspace while simultaneously preserving certain characteristics of the original domain data. The shared subspace can be manually defined or learned with an optimization framework. Some task-specific LST approaches [21, 52–54, 67] explicitly define a set of domain-independent meta-features that are common to both the domains. In general, defining the shared meta-features requires domain expertise. Predominantly, these meta-features are derived from structural, spatial, functional or temporal relationships between the original features. The manual construction of the common subspace is not optimal as domain-dependent features get ignored. Moreover, for most transfer scenarios, it is hard to determine the shared meta-features from heterogeneous feature spaces. Wang et al. [72–75] proposed several LST approaches that make the manifold assumption while learning the shared subspace. These approaches mimic a dimensionality reduction framework where the original geometric structure is preserved. The semi-supervised version of the manifold based LST approaches utilize few labeled target examples to match the conditional distributions in the shared manifold. Some of these approaches [72, 73] utilize a few cross-domain pairwise correspondences to improve the co-alignment of the transformed source and target domain data. However, for heterogeneous transfer tasks involving high-dimensional data, the manifold assumption does not hold well [80]. Pivots via non-negative joint matrix factorization (PJNMF) [91] is an unsupervised LST approach that collectively factorizes the heterogeneous feature spaces of the domains into latent domain-dependent features and domain-shared features (pivots). While factorizing, the latent features are kept orthogonal to each other so as to learn independent features. For cross-lingual transfer tasks, the domain-shared features are also known as topics. PJNMF is a task-specific approach as it is applicable to only those transfer tasks where the domains implicitly share a few key latent concepts such as cross-lingual text classification. Based on spectral embedding, Heterogeneous spectral Mapping (HeMap) [56] is designed as an unsupervised LST optimization framework that minimizes the marginal distribution differences between the transformed data of the domains in the latent space. As a consequence of constraining the projected target and source to be similar, sufficient cross-domain correspondences are required to learn robust transformations. HeMap adopts a matrix factorization solution, due to which out-of-sample data cannot be generalized. Heterogeneous Transfer Learning for Image Classification [92] is another LST optimization approach that leverages annotated tags on images as pivots to define correspondences for image and text modalities. Assuming some semantic relationship between the image and text data, the joint matrix factorization framework learns an enriched representation for image classification tasks. Semi-supervised subspace co-projection (SCP) [80] is a semi-supervised LST optimization approach that learns the target task in the shared subspace alongside with the mappings. In the shared subspace, the labeled examples are used to learn the task and match the conditional distributions whereas the unlabeled domain examples are used to match the marginal distributions. For high-dimensional datasets, the SCP approach requires large matrix inversions for its closed-form solution. In contrast to [59] which is an FSR based approach,

Co-regularized Heterogeneous Transfer Learning [78] is an LST approach that bridges heterogeneous domains with semantically related label spaces. The semi-supervised approach utilizes word2vec [44] embedding of the labels to determine the similarity between the labeled domain examples. Additionally, the Spatio-temporal similarities of the domain examples are used to define the closeness in the shared subspace. Unlike [59], the approach does not consider pushing dissimilar instances away. Heterogeneous Feature Augmentation (HFA) [18, 38] is a semi-supervised SVM-based optimization framework that uses an augmented feature space as the shared subspace for adaptation. The augmented feature space is derived from the original source and target features. As HFA does not explicitly output the mappings for the source and target domain, unseen target data cannot be generalized.

In contrast to the FSR approaches that directly transform the data from one domain to other, the LST approaches rely on the implicit feature relationships between the domains to determine a shared subspace where the data distributions are aligned. Determining the optimal dimension size of the shared subspace requires a grid search, which is an expensive task. Secondly, the solution for most LST optimization frameworks involves either an eigen decomposition problem [47, 56, 72] or a matrix factorization [91, 92] one. Consequently, these approaches suffer from the out-of-sample extension problem. Moreover, some LST approaches need a small number of correspondences [72, 76] to force the transformed data to be similar in the shared subspace. However, LST approaches estimate fewer parameters than their FSR counterparts.

2.4 Summary and Future Research

This chapter covers recent state-of-the-art generic approaches for DA and HDA. We also briefly describe a few motivating applications that require adaptation. [13, 35, 46, 50, 63, 69, 71, 87] and [17, 88] are other surveys that cover shallow DA and HDA approaches respectively in more detail. The focus of the present adaptation research has shifted more towards deep-neural networks as it is much easier to learn non-linear transformations. Most of the recent deep adaptation research can be viewed as a derivative of the prior shallow DA work. Nonetheless, the shallow DA methods are still preferred over their deep counterparts in most real-world online scenarios when there are limited compute resources.

The field of DA is still growing and there is a lot of scope for future research. A common assumption among transfer learning approaches is that there is abundant labeled data in the source domain. There is very limited research when only unlabeled data is available from an auxiliary domain(s). One major problem for performing knowledge transfer in the real-world tasks occurs when the target domain inherently has skewed data. In the unsupervised domain adaptation setting, handling the class imbalance problem is a major challenge. Prior approaches have utilized the pseudo-labels of the target domain examples to foster the alignment in the common space. However, such shallow DA/HDA approaches ignore the correctness of these pseudo-labels. A common assumption among the DA approaches is that the domains are related to some extent so as to avoid Negative-Transfer [55]. Unlike DA where it is easier to determine the relatedness of the domains as they have a common feature space, the future work can focus on finding a measure of domain similarity in the multi-source HDA setting in a task-independent manner. Secondly, most of the adaptation work focuses on the transfer setting where the source and target domain labels are same i.e. $Y_s = Y_t$ or the domains share a subset of domain labels $Y_s \cap Y_t \neq \phi$, another area to explore for knowledge transfer is when label spaces are not overlapping.

References

1. Aljundi, R., Emonet, R., Muselet, D., Sebban, M.: Landmarks-based kernelized subspace alignment for unsupervised domain adaptation. In: Proceedings of the IEEE Conference on Computer Vision and Pattern Recognition, pp. 56–63 (2015)
2. Al-Stouhi, S., Reddy, C.K.: Adaptive boosting for transfer learning using dynamic updates. In: Joint European Conference on Machine Learning and Knowledge Discovery in Databases, pp. 60–75. Springer, Berlin (2011)
3. Aytar, Y., Zisserman, A.: Tabula rasa: model transfer for object category detection. In: Proceedings of the IEEE International Conference on Computer Vision (ICCV), pp. 2252–2259 (2011)
4. Baktashmotlagh, M., Harandi, M.T., Lovell, B.C., Salzmann, M.: Unsupervised domain adaptation by domain invariant projection. In: Proceedings of the IEEE International Conference on Computer Vision, pp. 769–776 (2013)
5. Baktashmotlagh, M., Harandi, M.T., Lovell, B.C., Salzmann, M.: Domain adaptation on the statistical manifold. In: Proceedings of the IEEE Conference on Computer Vision and Pattern Recognition, pp. 2481–2488 (2014)
6. Belkin, M., Niyogi, P., Sindhwani, V.: Manifold regularization: a geometric framework for learning from labeled and unlabeled examples. J. Mach. Learn. Res. 7, 2399–2434 (2006)
7. Ben-David, S., Blitzer, J., Crammer, K., Pereira, F.: Analysis of representations for domain adaptation. In: Advances in Neural Information Processing Systems, pp. 137–144 (2007)
8. Blitzer, J., McDonald, R., Pereira, F.: Domain adaptation with structural correspondence learning. In: Proceedings of the 2006 Conference on Empirical Methods in Natural Language Processing, pp. 120–128 (2006)
9. Borchani, H., Varando, G., Bielza, C., Larrañaga, P.: A survey on multi-output regression. Wiley Interdiscip. Rev. Data Min. Knowl. Disc. 5(5), 216–233 (2015)
10. Borgwardt, K.M., Gretton, A., Rasch, M.J., Kriegel, H.P., Schölkopf, B., Smola, A.J.: Integrating structured biological data by kernel maximum mean discrepancy. Bioinformatics 22(14), e49–e57 (2006)
11. Breiman, L.: Random forests. Mach. Learn. 45, 5–32 (2001)
12. Bruzzone, L., Marconcini, M.: Domain adaptation problems: a DASVM classification technique and a circular validation strategy. IEEE Trans. Pattern Anal. Mach. Intell. 32(5), 770–787 (2010)
13. Csurka, G.: A comprehensive survey on domain adaptation for visual applications. In: Domain Adaptation in Computer Vision Applications, pp. 1–35. Springer, Berlin (2017)
14. Dai, W., Yang, Q., Xue, G.R., Yu, Y.: Boosting for transfer learning. In: Proceedings of the 24th International Conference on Machine Learning, pp. 193–200 (2007)

15. Daume III, H.: Frustratingly easy domain adaptation. In: Proceedings of the 45th Annual Meeting of the Association of Computational Linguistics, pp. 256–263 (2007)
16. Daumé III, H., Kumar, A., Saha, A.: Frustratingly easy semi-supervised domain adaptation. In: Proceedings of the 2010 Workshop on Domain Adaptation for Natural Language Processing, pp. 53–59. Association for Computational Linguistics, Stroudsburg (2010)
17. Day, O., Khoshgoftaar, T.M.: A survey on heterogeneous transfer learning. J. Big Data 4(1), 29 (2017)
18. Duan, L., Xu, D., Tsang, I.: Learning with augmented features for heterogeneous domain adaptation (2012). Preprint. arXiv:1206.4660
19. Duan, L., Xu, D., Tsang, I.W.H.: Domain adaptation from multiple sources: a domain-dependent regularization approach. IEEE Trans. Neural Netw. Learn. Syst. 23(3), 504–518 (2012)
20. Fernando, B., Habrard, A., Sebban, M., Tuytelaars, T.: Unsupervised visual domain adaptation using subspace alignment. In: Proceedings of the IEEE International Conference on Computer Vision, pp. 2960–2967 (2013)
21. Feuz, K.D.: Preparing smart environments for life in the wild: feature-space and multi-view heterogeneous learning. Ph.D. Thesis, Washington State University (2014)
22. Feuz, K.D., Cook, D.J.: Heterogeneous transfer learning for activity recognition using heuristic search techniques. Int. J. Pervasive Comput. Commun. 10, 393–418 (2014)
23. Freund, Y., Schapire, R.E.: A decision-theoretic generalization of on-line learning and an application to boosting. J. Comput. Syst. Sci. 55(1), 119–139 (1997)
24. Glorot, X., Bordes, A., Bengio, Y.: Domain adaptation for large-scale sentiment classification: a deep learning approach. In: Proceedings of the 28th International Conference on Machine Learning (ICML-11), pp. 513–520 (2011)
25. Gong, B., Shi, Y., Sha, F., Grauman, K.: Geodesic flow kernel for unsupervised domain adaptation. In: Proceedings of the IEEE Conference on Computer Vision and Pattern Recognition (CVPR) (2012)
26. Gong, B., Grauman, K., Sha, F.: Connecting the dots with landmarks: discriminatively learning domain-invariant features for unsupervised domain adaptation. In: Proceedings of the ACM International Conference on Machine Learning (ICML), pp. 222–230 (2013)
27. Gopalan, R., Li, R., Chellappa, R.: Domain adaptation for object recognition: an unsupervised approach. In: Proceedings of the IEEE Internatinal Conference on Computer Vision (ICCV), pp. 999–1006. IEEE, Piscataway (2011)
28. Gopalan, R., Li, R., Chellappa, R.: Unsupervised adaptation across domain shifts by generating intermediate data representations. IEEE Trans. Pattern Anal. Mach. Intell. 36(11), 2288–2302 (2014)
29. Gretton, A., Smola, A., Huang, J., Schmittfull, M., Borgwardt, K., Schölkopf, B.: Covariate shift by kernel mean matching. Dataset Shift Mach. Learn. 3(4), 5 (2009)
30. He, J., Liu, Y., Yang, Q.: Linking heterogeneous input spaces with pivots for multi-task learning. In: Proceedings of the SIAM International Conference on Data Mining, pp. 181–189 (2014)
31. Hoffman, J., Rodner, E., Donahue, J., Darrell, T., Saenko, K.: Efficient learning of domain-invariant image representations. In: International Conference on Learning Representations (2013)
32. Hu, D.H., Yang, Q.: Transfer learning for activity recognition via sensor mapping. In: Proceedings of the 22nd International Joint Conference on Artificial Intelligence, pp. 1962–1967 (2011)
33. Huang, J., Gretton, A., Borgwardt, K., Schölkopf, B., Smola, A.J.: Correcting sample selection bias by unlabeled data. In: Advances in Neural Information Processing Systems, pp. 601–608 (2007)
34. Hubert Tsai, Y.H., Yeh, Y.R., Frank Wang, Y.C.: Learning cross-domain landmarks for heterogeneous domain adaptation. In: Proceedings of the IEEE Conference on Computer Vision and Pattern Recognition, pp. 5081–5090 (2016)

35. Jiang, J.: A literature survey on domain adaptation of statistical classifiers. Comput. Sci. **3**, 1–12 (2008). http://sifaka.cs.uiuc.edu/jiang4/domainadaptation/survey
36. Kanamori, T., Hido, S., Sugiyama, M.: Efficient direct density ratio estimation for non-stationarity adaptation and outlier detection. In: Advances in Neural Information Processing Systems, pp. 809–816 (2009)
37. Kanamori, T., Hido, S., Sugiyama, M.: A least-squares approach to direct importance estimation. J. Mach. Learn. Res. **10**, 1391–1445 (2009)
38. Li, W., Duan, L., Xu, D., Tsang, I.W.: Learning with augmented features for supervised and semi-supervised heterogeneous domain adaptation. IEEE Trans. Pattern Anal. Mach. Intell. **36**, 1134–1148 (2014)
39. Li, Y., Yang, M., Zhang, Z.M.: A survey of multi-view representation learning. IEEE Trans. Knowl. Data Eng. **31**, 1863–1883 (2018)
40. Long, M., Wang, J., Ding, G., Sun, J., Yu, P.S.: Transfer feature learning with joint distribution adaptation. In: Proceedings of the ACM International Conference on Machine Learning (ICML), pp. 2200–2207 (2013)
41. Long, M., Wang, J., Ding, G., Pan, S.J., Philip, S.Y.: Adaptation regularization: a general framework for transfer learning. IEEE Trans. Knowl. Data Eng. **26**(5), 1076–1089 (2014)
42. Long, M., Wang, J., Ding, G., Sun, J., Yu, P.S.: Transfer joint matching for unsupervised domain adaptation. In: Proceedings of the IEEE Conference on Computer Vision and Pattern Recognition (CVPR) (2014)
43. Matasci, G., Volpi, M., Kanevski, M., Bruzzone, L., Tuia, D.: Semisupervised transfer component analysis for domain adaptation in remote sensing image classification. IEEE Trans. Geosci. Remote Sens. **53**(7), 3550–3564 (2015)
44. Mikolov, T., Sutskever, I., Chen, K., Corrado, G.S., Dean, J.: Distributed representations of words and phrases and their compositionality. In: Advances in Neural Information Processing Systems, pp. 3111–3119 (2013)
45. Pan, J.: Feature based transfer learning with real-world applications. Ph.D. Thesis, Hong Kong University of Science and Technology (2010)
46. Pan, S.J., Yang, Q.: A survey on transfer learning. IEEE Trans. Knowl. Data Eng. **22**, 1345–1359 (2010)
47. Pan, S.J., Kwok, J.T., Yang, Q.: Transfer learning via dimensionality reduction. In: Proceedings of the 23rd National Conference on Artificial Intelligence, pp. 677–682. AAAI, Menlo Park (2008)
48. Pan, S.J., Tsang, I.W., Kwok, J.T., Yang, Q.: Domain adaptation via transfer component analysis. In: Proceedings of the 21st International Joint Conference on Artificial Intelligence, pp. 1187–1192 (2009)
49. Pan, S.J., Ni, X., Sun, J.T., Yang, Q., Chen, Z.: Cross-domain sentiment classification via spectral feature alignment. In: Proceedings of the 19th International Conference on World Wide Web, pp. 751–760. ACM, New York (2010)
50. Patel, V.M., Gopalan, R., Li, R., Chellappa, R.: Visual domain adaptation: a survey of recent advances. IEEE Signal Process. Mag. **32**(3), 53–69 (2015)
51. Prettenhofer, P., Stein, B.: Cross-language text classification using structural correspondence learning. In: Proceedings of the 48th Annual Meeting of the Association for Computational Linguistics, ACL '10, pp. 1118–1127 (2010)
52. Qi, G.J., Aggarwal, C., Huang, T.: Transfer learning of distance metrics by cross-domain metric sampling across heterogeneous spaces. In: Proceedings of the 2012 SIAM International Conference on Data Mining, pp. 528–539. SIAM, Philadelphia (2012)
53. Raina, R., Battle, A., Lee, H., Packer, B., Ng, A.Y.: Self-taught learning: transfer learning from unlabelled data. In: Proceedings of the 24th International Conference on Machine Learning, pp. 759–766 (2007)
54. Rashidi, P., Cook, D.J.: Multi home transfer learning for resident activity discovery and recognition. In: Proceedings of the International Workshop on Knowledge Discovery from Sensor Data, pp. 56–63 (2010)

55. Rosenstein, M.T., Marx, Z., Kaelbling, L.P., Dietterich, T.G.: To transfer or not to transfer. In: NIPS 2005 Workshop on Transfer Learning, vol. 898 (2005)
56. Shi, X., Yu, P.: Dimensionality reduction on heterogeneous feature space. In: Proceedings of the 12th IEEE International Conference on Data Mining, pp. 635–644 (2012)
57. Si, S., Tao, D., Geng, B.: Bregman divergence-based regularization for transfer subspace learning. IEEE Trans. Knowl. Data Eng. **22**(7), 929–942 (2010)
58. Sugiyama, M., Nakajima, S., Kashima, H., Buenau, P.V., Kawanabe, M.: Direct importance estimation with model selection and its application to covariate shift adaptation. In: Advances in Neural Information Processing Systems (NIPS), pp. 1433–1440 (2008)
59. Sukhija, S., Krishnan, N.C.: Web-induced heterogeneous transfer learning with sample selection. In: Joint European Conference on Machine Learning and Knowledge Discovery in Databases, pp. 777–793. Springer, Berlin (2018)
60. Sukhija, S., Krishnan, N.C.: Supervised heterogeneous feature transfer via random forests. Artif. Intell. **268**, 30–53 (2019)
61. Sukhija, S., Krishnan, N.C., Singh, G.: Supervised heterogeneous domain adaptation via random forests. In: Proceedings of the 25th International Joint Conference on Artificial Intelligence, pp. 193–200 (2016)
62. Sukhija, S., Krishnan, N.C., Kumar, D.: Supervised heterogeneous transfer learning using random forests. In: Proceedings of the ACM India Joint International Conference on Data Science and Management of Data, pp. 157–166 (2018)
63. Sun, S., Shi, H., Wu, Y.: A survey of multi-source domain adaptation. Inf. Fusion **24**, 84–92 (2015)
64. Sun, B., Feng, J., Saenko, K.: Return of frustratingly easy domain adaptation. In: Thirtieth AAAI Conference on Artificial Intelligence (2016)
65. Tommasi, T., Orabona, F., Caputo, B.: Safety in numbers: Learning categories from few examples with multi model knowledge transfer. In: 2010 IEEE Computer Society Conference on Computer Vision and Pattern Recognition, pp. 3081–3088. IEEE, Piscataway (2010)
66. Torralba, A., Efros, A.A.: Unbiased look at dataset bias. In: Proceedings of the IEEE Conference on Computer Vision and Pattern Recognition (CVPR), pp. 1521–1528 (2011)
67. van Kasteren, T.L.M., Englebienne, G., Kröse, B.J.A.: Transferring knowledge of activity recognition across sensor networks. In: Proceedings of the 8th International Conference on Pervasive Computing, pp. 283–300 (2010)
68. Vapnik, V., Vapnik, V.: Statistical Learning Theory, pp. 156–160. Wiley, New York (1998)
69. Venkateswara, H., Chakraborty, S., Panchanathan, S.: Deep-learning systems for domain adaptation in computer vision: Learning transferable feature representations. IEEE Signal Process. Mag. **34**(6), 117–129 (2017)
70. Wang, L.: Support Vector Machines: Theory and Applications, vol. 177. Springer Science & Business Media, Berlin (2005)
71. Wang, M., Deng, W.: Deep visual domain adaptation: a survey. Neurocomputing **312**, 135–153 (2018)
72. Wang, C., Mahadevan, S.: Manifold alignment using procrustes analysis. In: Proceedings of the International Conference on Machine Learning, pp. 1120–1127 (2008)
73. Wang, C., Mahadevan, S.: A general framework for manifold alignment. In: AAAI Fall Symposium: Manifold Learning and Its Applications (2009)
74. Wang, C., Mahadevan, S.: Manifold alignment without correspondence. In: Proceedings of the 21st International Joint Conference on Artificial Intelligence, pp. 1273–1278 (2009)
75. Wang, C., Mahadevan, S.: Heterogeneous domain adaptation using manifold alignment. In: IJCAI Proceedings-International Joint Conference on Artificial Intelligence (2011)
76. Wang, C., Mahadevan, S.: Manifold alignment preserving global geometry. In: Proceedings of the Twenty-Third International Joint Conference on Artificial Intelligence (2013)
77. Wei, Y., Zheng, Y., Yang, Q.: Transfer knowledge between cities. In: Proceedings of the 22nd ACM SIGKDD International Conference on Knowledge Discovery and Data Mining, pp. 1905–1914 (2016)

78. Wei, Y., Zhu, Y., Leung, C.W.k., Song, Y., Yang, Q.: Instilling social to physical: co-regularized heterogeneous transfer learning. In: Proceedings of the AAAI National Conference on Artificial Intelligence, pp. 1338–1344 (2016)
79. Wu, P., Dietterich, T.G.: Improving SVM accuracy by training on auxiliary data sources. In: Proceedings of the Twenty-First International Conference on Machine Learning, ICML '04. ACM, New York (2004). http://doi.acm.org/10.1145/1015330.1015436
80. Xiao, M., Guo, Y.: Machine Learning and Knowledge Discovery in Databases: European Conference, ECML PKDD 2015, Porto, September 7–11. In: Proceedings, Part II, chap. Semi-supervised Subspace Co-projection for Multi-class Heterogeneous Domain Adaptation, pp. 525–540. Springer International Publishing, Basel (2015)
81. Yang, J., Yan, R., Hauptmann, A.G.: Adapting SVM classifiers to data with shifted distributions. In: Seventh IEEE International Conference on Data Mining Workshops. ICDM Workshops 2007, pp. 69–76 (2007)
82. Yang, J., Yan, R., Hauptmann, A.G.: Cross-domain video concept detection using adaptive SVMs. In: Proceedings of the ACM International Conference on Multimedia (ACM-MM), pp. 188–197 (2007)
83. Yang, Q., Pan, S.J., Zheng, V.W.: Estimating location using Wi-Fi. IEEE Intell. Syst. 23(1), 8–13 (2008)
84. Yang, Q., Chen, Y., Xue, G.R., Dai, W., Yu, Y.: Heterogeneous transfer learning for image clustering via the social web. In: Proceedings of the Joint Conference of the 47th Annual Meeting of the ACL and the 4th International Joint Conference on Natural Language Processing of the AFNLP, pp. 1–9. Association for Computational Linguistics, Stroudsburg (2009)
85. Yao, Y., Doretto, G.: Boosting for transfer learning with multiple sources. In: 2010 IEEE Computer Society Conference on Computer Vision and Pattern Recognition, pp. 1855–1862. IEEE, Piscataway (2010)
86. Zadrozny, B.: Learning and evaluating classifiers under sample selection bias. In: Proceedings of the ACM International Conference on Machine Learning (ICML), p. 114 (2004)
87. Zhang, L.: Transfer adaptation learning: a decade survey (2019). Preprint. arXiv:1903.04687
88. Zhang, J., Li, W., Ogunbona, P.: Transfer learning for cross-dataset recognition: a survey (2017). Preprint. arXiv:1705.04396
89. Zhong, E., Fan, W., Peng, J., Zhang, K., Ren, J., Turaga, D., Verscheure, O.: Cross domain distribution adaptation via kernel mapping. In: Proceedings of the 15th ACM SIGKDD International Conference on Knowledge Discovery and Data Mining, pp. 1027–1036. ACM, New York (2009)
90. Zhou, J.T., Tsang, I.W., Pan, S.J., Tan, M.: Heterogeneous domain adaptation for multiple classes. In: Proceedings of the 17th International Conference on Artificial Intelligence and Statistics, pp. 1095–1103 (2014)
91. Zhou, G., He, T., Wu, W., Hu, X.T.: Linking heterogeneous input features with pivots for domain adaptation. In: Proceedings of the 24th International Conference on Artificial Intelligence, IJCAI'15, pp. 1419–1425. AAAI Press, Palo Alto (2015)
92. Zhu, Y., Chen, Y., Lu, Z., Pan, S.J., Xue, G.R., Yu, Y., Yang, Q.: Heterogeneous transfer learning for image classification. In: Proceedings of the AAAI National Conference on Artificial Intelligence, pp. 1304–1310 (2011)

Part II
Domain Alignment in the Feature Space

Chapter 3
d-SNE: Domain Adaptation Using Stochastic Neighborhood Embedding

Xiong Zhou, Xiang Xu, Ragav Venkatesan, Gurumurthy Swaminathan, and Orchid Majumder

3.1 Introduction

Thanks to large-scale datasets like ImageNet [2], fine-tuning a pretrained model allows us to achieve some unprecedented results on relatively small-scale datasets. Due to the domain shift problem, however, it is hard to repurpose the pretrained model to work in the new domain when the labeled data are extremely limited. Domain adaptation, on the other hand, relies on the class relationship to transfer knowledge from one domain to another.

Existing domain adaptation techniques can be broadly grouped into two categories: domain transformation and feature transformation. In domain transformation methods, data from the source domain are transformed into the target domain or vice-versa, so that the labeled data from the source domain can be used to train a unified classifier. Such transformation is usually done by using a generative adversarial network (GAN) [7]. Liu and Tuzel [14] proposed to synthesize images in both source and target domains with a pair of GANs. A joint distribution of two domains was learned by enforcing a weight-sharing constraint. Rather than generating images from both domains, Bousmalis et al. [1] introduced a method to transform images from the source domain to the target domain and forced them to be similar with a GAN discriminator. Similarly, Hoffman et al. [11] used the CycleGAN [34] to transfer the style of source images, while preserving the content with a semantic consistency loss. Unlike the previous methods tried to transform data from one domain to another, Russo et al. [21] considered a bi-directional image transformations that united the source and target domain.

X. Zhou (✉) · X. Xu · R. Venkatesan · G. Swaminathan · O. Majumder
AWS AI, Seattle, WA, USA
e-mail: xiongzho@amazon.com; xiangx@amazon.com; ragavven@amazon.com; gurumurs@amazon.com; orchid@amazon.com

© Springer Nature Switzerland AG 2020   43
H. Venkateswara, S. Panchanathan (eds.), *Domain Adaptation in Computer Vision with Deep Learning*, https://doi.org/10.1007/978-3-030-45529-3_3

Feature transformation-based methods aim to extract features that are invariant across domains. Ganin et al. [5] introduced the domain adversarial training for domain adaptation, where the gradients for domain classifier was intentionally flipped before back-propagating. Features were considered as domain-invariant when the domain classifier could not decide their origins. Such a framework was generalized by Tzeng et al. in [29]. Another line of research is to form the domain alignment and classification as a multi-task learning problem [3, 8, 20, 25, 28], where they tried extract features that are discriminative as well as domain-invariant.

In this chapter, we introduce a novel domain adaptation algorithm that only requires a few labeled samples from the target domain. The proposed method creates a latent space using stochastic neighborhood embedding, classes from two domains are aligned based on a modified Hausdorff distance metric. The proposed idea can be used as a loss to train neural network in an end-to-end fashion. Extensive experiments demonstrate our algorithm is robust and outperforms the state-of-the-art in multiple benchmark datasets. We also show that the proposed method has good capability in domain generalization, which is typically not the case in conventional domain adaptation approaches. This chapter is partially based on earlier work from [32].

3.2 Proposed Method

Denote the source domain as $\mathcal{D}^s = \{(x_i^s, y_i^s)\}_{i=1}^{N^s}$ and the target domain as $\mathcal{D}^t = \{(x_j^t, y_j^t)\}_{j=1}^{N^t}$. In a domain adaptation problem, the target domain typically has much less labeled samples than the source domain, $N^t << N^s$. We also consider that the label spaces are the same for both domains, i.e., $\{y^s, y^t\} \in [0, 1, \ldots c - 1]$. The distance between a source sample and a target sample in the transformed space can be defined as:

$$d(x_i^s, x_j^t) = \|\Phi_{\mathcal{D}^s}(x_i^s) - \Phi_{\mathcal{D}^t}(x_j^t)\|^2, \tag{3.1}$$

where $\Phi_{\mathcal{D}^s}(\cdot) \rightarrow \mathbb{R}^d$ and $\Phi_{\mathcal{D}^t}(\cdot) \rightarrow \mathbb{R}^d$ are neural networks, parameterized by w_s and w_t respectively, that transform samples from both domains to a common latent space. Inspired by the stochastic neighborhood embedding (SNE) [10], we define the probability that the target sample $x_j^t \in \mathcal{D}^t$ select the source sample $x_i^s \in \mathcal{D}^s$ as its neighbor as follows,

$$p_{ij} = \frac{\exp(-d(x_i^s, x_j^t))}{\sum_{x \in \mathcal{D}^s} \exp(-d(x, x_j^t))}. \tag{3.2}$$

Knowing labels for both x_i^s and x_j^t, we want to maximize p_{ij} if $y_i^s = y_j^t$, otherwise minimize it. It is worth noting that training samples in the source domain are selected based on a probability distribution that favors nearby points over faraway ones. In other words, the framework operates in a local manner, in which the larger the distance between x_i^s and x_j^t, the smaller probability for selecting x_i^s as the neighbor of x_j^t. Such locality has proven to be useful when dealing with multimodal data [30].

Under this stochastic selection rule, the probability of finding the right neighbor for x_j^t is

$$p_j = \frac{\sum_{x \in \mathcal{D}_k^s} \exp(-d(x, x_j^t))}{\sum_{x \in \mathcal{D}^s} \exp(-d(x, x_j^t))} = \sum_{i=0}^{N_k^s} p_{ij}, \qquad (3.3)$$

where $\mathcal{D}_k^s = \{\forall x_i^s | y_i^s = k\}$ is a set of source samples that has the same label as y_j^t and $N_k^s = |\mathcal{D}_k^s|$.

Given a target sample x_j^t with its label $y_j^t = k$, the source domain can be split into a *same-class set* $\mathcal{D}_k^s$ and a *different-class set*, $\mathcal{D}_{k'}^s$. The denominator in Eq. (3.3) can now be decomposed as $\sum_{x \in \mathcal{D}_k^s} \exp(-d(x, x_j^t)) + \sum_{x \in \mathcal{D}_{k'}^s} \exp(-d(x, x_j^t))$. Maximizing the probability p_j is equivalent to minimizing $\frac{1}{p_j}$, which in fact aligns the target domain with the source domain class by class. Therefore, the loss of x_j^t for the domain adaptation problem can be derived as:

$$\sum_{x_j \in \mathcal{D}^t} \frac{1}{p_j} = \frac{\sum_{x \in \mathcal{D}_{k'}^s} \exp(-d(x, x_j))}{\sum_{x \in \mathcal{D}_k^s} \exp(-d(x, x_j))}, \quad \text{for } k = y_j. \qquad (3.4)$$

Minimizing $\log \frac{1}{p_j}$ can be considered as minimizing the ratio of intra-class distances to inter-class distances in the latent space.

$$\mathcal{L} = \log \left(\frac{\sum_{x \in \mathcal{D}_{k'}^s} \exp(-d(x, x_j))}{\sum_{x \in \mathcal{D}_k^s} \exp(-d(x, x_j))}, \quad \text{for } k = y_j \right). \qquad (3.5)$$

In practice, we have neural networks, $\Phi_{\mathcal{D}^s}$ and $\Phi_{\mathcal{D}^t}$, as feature extractors, the sum of exponentials in Eq. (3.5) may cause unstable training due to scaling problem. To address this issue, we use a modified Hausdorff distance to relax the likelihood. Instead of optimizing the distances in the entire set, we only minimize the largest distance between samples of the same class and maximize the smallest distance between the samples of different classes. The final loss for domain adaptation can be written as:

$$\tilde{\mathcal{L}} = \sup_{x \in \mathcal{D}_k^s} \{a | a \in d(x, x_j)\} - \inf_{x \in \mathcal{D}_{k'}^s} \{b | b \in d(x, x_j)\}, \quad \text{for } k = y_j. \qquad (3.6)$$

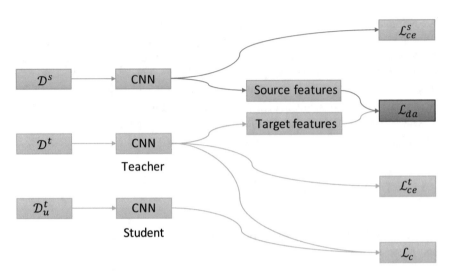

Fig. 3.1 The learning setup. The segment in the bottom in lighter shade and dotted lines is the semi-supervised extension

Figure 3.1 illustrates our setup for the end to end training. The feature extractors are two independent neural networks $\Phi_{\mathcal{D}^s}$ and $\Phi_{\mathcal{D}^t}$. If the input data from the source and target domains have the same dimensionality, a single network can be shared by the two domains. With the domain adaptation loss $\tilde{\mathcal{L}}$, d-SNE transfers the supervision from the source domain to the target domain by selecting the right neighbors for the target samples. Notice that we have labels in both domains, standard cross-entropy losses, $\mathcal{L}_{ce}^s$ and $\mathcal{L}_{ce}^t$, are also used to train the networks. Training with these classification losses not only leads to more discriminative features but also proved to be helpful for domain alignment [33]. Our learning formulation is therefore defined by,

$$\underset{w_s, w_d}{\arg\min} \tilde{\mathcal{L}} + \alpha \mathcal{L}_{ce}^s + \beta \mathcal{L}_{ce}^t. \tag{3.7}$$

The weighting factors α and β can be tuned to balance different losses based on datasets. In this work, we apply equal weights to the source and target domains. The aforementioned losses are all focusing on labeled data, while having access to unlabeled data can also boost the model performance. We introduce an easy way to extend the framework into a semi-supervised setting to accommodate unlabeled data. Consider that the unlabeled data from the target domain is $\mathcal{D}_u^t$. Following the Mean-Teacher model proposed by Tarvainen et al., [27], we make a copy of the target network using its exponential moving average. For each unlabeled target sample, a pair of inputs is created by applying random data augmentations, such as flipping, cropping, color jittering, etc. These two sibling samples are then used to train the target network $\Phi_{\mathcal{D}^t}$ and its copy $\hat{\Phi}_{\mathcal{D}^t}$. Remember that there is no label for these input pairs, we cannot train the network with a normal cross-entropy loss. However, since the input pair contains two variants of the same sample and belongs to the same class, we can enforce the networks to predict the same label by using a consistency loss. This is done by taking an L_2 loss over the embeddings of input pair. The semi-supervised extension is depicted as the bottom row in Fig. 3.1.

3.3   Experiments and Results

In this section, we evaluate the proposed *d*-SNE on three datasets: (1) Digits dataset. It contains four domains as MNIST [13], MNIST-M [5], SVHN [18], and USPS [12]. MNIST has 70,000 grayscale images with size 28×28, while MNIST-M is variant of MNIST where the background is replaced by random patches from color images. SVHN contains 99,289 RGB images that are house numbers captured in Google Street View. USPS has 9298 images that are scanned from envelopes by the U.S. Postal Service. (2) Office-31 dataset [22]. It has three domains of images of office items that are collected in different conditions. The domain A has 2817 images, which is collected from the Amazon website, domain D has 498 images that are captured by DSLR camera, and domain W has 795 images are captured by web camera. (3) VisDA-C dataset [19]. It includes a synthetic domain with 152,397 images and a real domain with 4000 images. Examples from these datasets are shown in Fig. 3.2. Our algorithm code and experimental setup can be found at https://github.com/aws-samples/d-sne-domain-adaptation.

3.3.1   Digit Dataset

In the first set of experiments, we compare the proposed *d*-SNE with the state-of-the-art supervised domain adaptation approaches CCSA [16] and FADA [15]. Following the protocol in [16], a total of 2000 images are sampled from MNIST as the source domain. USPS is used as the target domain. Besides the labeled data from the source domain, a small number of samples per-class ranging from 1 to 7 are randomly selected from the target domain for training. To keep a fair comparison, we used the same network architecture as CCSA and FADA. Table 3.1 shows the overall classification accuracies for adaptation from MNIST to USPS. As you can see that our method consistently outperforms both CCSA and FADA in all testing cases. *d*-SNE achieved relatively large improvement when target training samples are extremely limited, e.g., $|\mathcal{D}_k^t| = 1$. For the non-adaptation case ($|\mathcal{D}_k^t| = 0$),

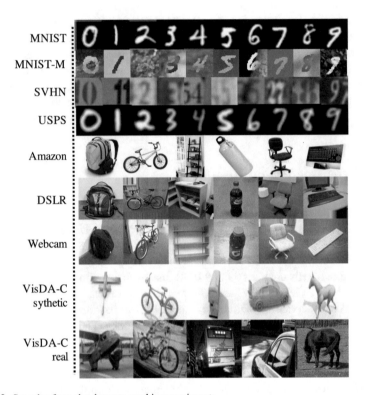

Fig. 3.2 Samples from the datasets used in experiments

Table 3.1 MNIST → USPS datasets

$|\mathcal{D}_k^t|$, $\forall k$	0	1	3	5	7
CCSA [16]	65.40	85.00	90.10	92.40	92.90
FADA [15]	65.40	89.10	91.90	93.40	94.40
d-SNE	73.01	**92.90**	**93.55**	**95.13**	**96.13**

$|\mathcal{D}_k^t|$, $\forall k$ is essentially number of samples per-class from the target domain. As can be seen, d-SNE is clearly able to outperform the states-of-the-art in all scenarios. As the cardinality of the samples per-class increases, the performance across the algorithms converge

the model was trained with only source data and directly applied to the target domain. It can be noticed that even though we used the same architecture, our implementation obtained higher accuracy than CCSA and FADA. We attribute this to a better hyperparameter tuning. We were not able to out-tune their parameters with both their and our implementations for other four cases, therefore, we consider the reported numbers as the best for CCSA and FADA.

In the second set of experiments, we used all four digit datasets to create five domain adaptation scenarios: MNIST → MNIST-M, MNIST ↔ USPS, and MNIST ↔ SVHN. Several domain adaptation algorithms, including both supervised and unsupervised, are used for comparison. Unsupervised methods use all unlabeled samples from the target domain, while supervised ones only use 10 labeled samples per class from the target domain for training. Since multiple different architectures were used in these baselines, it is hard for us to keep a fair comparison with every method. In this set of experiments, our base network was a modified LeNet introduced in [31]. The overall classification accuracies are shown in Table 3.2. Our *d*-SNE achieved better performance compared to the supervised baselines in all cases. In domain pairs like MNIST → MNIST-M and SVHN → MNIST, unsupervised approaches in general obtained higher accuracies than supervised approaches. The proposed *d*-SNE even outperformed unsupervised methods in some cases like MNIST↔ USPS despite the fact that we use much less training data. We suspect that might be due to the relatively lower intra-class variance in MNIST and USPS compared to datasets like MNIST-M and SVHN. With only a few samples from the target domain, supervised method suffered to catch such variance. While unsupervised methods were able to handle the relatively high intra-class variance with the help of additional unlabeled data from the target domain. The performance of our method was further improved by adding the semi-supervised branch to the network. Although the comparison between our method and unsupervised methods is unfair, we believe in general the semi-supervised extension can enhance the *d*-SNE and allow us to consume unlabeled data. Figure 3.3 shows the t-SNE visualization for all adaptation cases. As can be seen that the cluster structure is better after the domain adaptation.

Since we train the network with data from both domains with a alignment loss, we expect the model to work well on not only the target domain but also the source domain. We refer to this additional benefit of *d*-SNE as domain generalization. To validate this hypothesis, we measure the model's accuracy on the source data without re-training or fine-tuning after the domain adaptation. Accuracies before and after adaptation in the source domain are shown in Table 3.3. We can notice that the model performance is actually improved in the source domain by doing domain adaptation.

3.3.2 Office Dataset

Office dataset [22] is a small-scale dataset that contains 3 domains of images. For our experiments, we followed the protocol used in [15, 16, 28]. In particular, we randomly selected 20 samples per class from domain A, 8 samples per class from domain D and domain W as source data. For target data, 3 samples were selected

Table 3.2 Classification accuracy for domain adaptation methods on digits datasets

Method	$|\mathcal{D}_k^t|, \forall k$	Setting	MNIST → MNIST-M	MNIST → USPS	USPS → MNIST	MNIST → SVHN	SVHN → MNIST
PixelDA [1]		$\mathcal{U}$	98.20	95.90	–	–	–
ADA [8]			87.47	–	–	–	97.60
I2I [17]			–	95.1	92.2	–	92.1
DIRT-T [25]			98.90	–	–	54.50	**99.40**
SE [4]			–	98.26 ± 0.11	98.07 ± 2.82	13.96 ± 4.41	99.18 ± 0.12
SBADA-GAN [21]			**99.40**	95.04	97.60	61.08	76.14
G2A [23]			–	95.30 ± 0.70	90.80 ± 1.30	–	92.40 ± 0.90
FADA [15]	7	$\mathcal{S}$	–	94.40	91.50	47.00	87.20
CCSA [16]	10		78.29 ± 2.00	97.27 ± 0.19	95.71 ± 0.42	37.63 ± 3.62	94.57 ± 0.40
d-SNE	0	$\mathcal{S}$	50.98 ± 1.64	93.16 ± 0.71	83.37 ± 0.93	26.22 ± 2.02	66.02 ± 0.72
	7		84.62 ± 0.04	97.53 ± 0.10	97.52 ± 0.08	53.19 ± 0.28	95.68 ± 0.03
	10		*87.80 ± 0.16*	**99.00 ± 0.08**	**98.49 ± 0.35**	**61.73 ± 0.47**	*96.45 ± 0.20*
d-SNE	10	$\mathcal{SS}$	94.12	–	–	77.63 ± 0.26	97.60

The unsupervised setting ($\mathcal{U}$) uses all the images in the target domain. The supervised setting ($\mathcal{S}$) uses 10 labeled samples per-class from the target domain. We reimplemented CCSA and FADA using the same network and settings as our method. The best results are marked in **bold**. If the best result is not in the supervised-only setting, we mark the best among the supervised-only methods in *italics*. The results are averaged over three runs and we report mean and standard deviations over the three runs

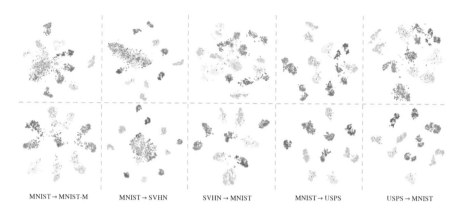

MNIST→MNIST-M   MNIST→SVHN   SVHN→MNIST   MNIST→USPS   USPS→MNIST

Fig. 3.3 t-SNE visualizations without (top) and with (bottom) domain adaptations

Table 3.3 Results of domain generalization

	MNIST → MNIST-M	MNIST → SVHN	SVHN → MNIST
Before	99.45%	99.51%	88.96%
After	99.51%	99.59%	94.94%

from all 3 domains. The rest of samples from the target domain were used for evaluation.

For the base network, we modified the ResNet-101 [9] by adding two extra dense layers to extract feature representations with 512 dimensions. Pre-trained weights from ImageNet were used as the initialization. Classification results of office dataset are reported in Table 3.4. Without using any unlabeled data from the target domain, *d*-SNE significantly outperformed all the baselines including unsupervised approaches. In cases like A↔W and A↔D , where the domain shift is relatively large, *d*-SNE showed larger margins compared to other methods. We attribute this improvement to the class by class alignment property of *d*-SNE. Since the supervised methods CCSA [16] and FADA [15] both used VGG-16 [26] as base networks, we also report the accuracy for *d*-SNE with a VGG-16 base network to keep the comparison fair. The proposed method still achieved higher accuracies than both CCSA and FADA in the majority of cases. It is worth noting that our non-adaptation baseline ($|\mathcal{D}_k^t| = 0$) with VGG-16 was actually worse than that in CCSA and FADA. That is to say, *d*-SNE achieved better performance with a worse base model, which highlights the strength of the proposed domain adaptation loss.

Table 3.4 Results of office31 experiments

Method	$|\mathcal{D}_k^t|, \forall k$	Setting	A → D	A → W	D → A	D → W	W → A	W → D	Avg.
DANN [5]		$\mathcal{U}$	–	73.00	–	96.40	–	99.20	–
DRCN [6]			67.10 ± 0.30	68.70 ± 0.30	56.00 ± 0.50	96.40 ± 0.30	54.09 ± 0.50	99.00 ± 0.2	73.60
kNN-Ad [24]			84.10	81.10	58.30	96.40	63.80	99.20	80.48
I2I [17]			71.10	75.30	50.10	96.50	52.10	99.60	74.12
G2A [23]			87.70 ± 0.50	89.50 ± 0.50	72.80 ± 0.30	97.90 ± 0.30	71.40 ± 0.40	99.8 ± 0.4	86.50
SDA [28]	3	$\mathcal{S}$	86.10 ± 1.20	82.70 ± 0.80	66.20 ± 0.30	95.70 ± 0.50	65.00 ± 0.5	97.60 ± 0.20	82.22
FADA [15]	3		88.20 ± 1.00	88.10 ± 1.20	68.10 ± 0.60	96.40 ± 0.80	71.10 ± 0.90	97.50 ± 0.90	84.90
CCSA [16]	0		61.20 ± 0.90	62.3 ± 0.80	58.5 ± 0.80	80.1 ± 0.60	51.6 ± 0.90	95.6 ± 0.70	68.20
CCSA [16]	3		89.00 ± 1.20	88.20 ± 1.00	71.80 ± 0.50	96.40 ± 0.80	72.10 ± 1.00	97.60 ± 0.40	85.80
d-SNE (VGG-16)	0	$\mathcal{S}$	62.40 ± 0.40	61.49 ± 0.75	48.92 ± 1.42	82.24 ± 1.42	47.52 ± 0.94	90.42 ± 1.00	65.49
	3		91.44 ± 0.23	90.13 ± 0.07	71.06 ± 0.18	97.10 ± 0.07	71.74 ± 0.42	97.46 ± 0.24	86.49
d-SNE (ResNet-101)	0	$\mathcal{S}$	80.41 ± 0.79	75.26 ± 1.32	67.39 ± 0.18	96.39 ± 0.41	65.55 ± 1.91	98.31 ± 1.87	80.55
	3		**94.65 ± 0.38**	**96.58 ± 0.14**	**75.51 ± 0.44**	**99.10 ± 0.24**	**74.20 ± 0.24**	**100.00 ± 0.00**	**90.01**

d-SNE with ResNet-101 base network achieves the best results with only 3 samples in the target domain, while *d*-SNE with VGG-16 base network outperforms the baselines in the majority of cases

Table 3.5 Results on the VisDA-C dataset

Method	Setting	$|\mathcal{D}_k^t| = 0,\ \forall k$	Adaptation
G2A [23]	$\mathcal{U}$	44.50	77.10
SE [4]	$\mathcal{SS}$	52.80	85.40
CCSA [16]	$\mathcal{S}$	52.80	76.89
d-SNE	$\mathcal{S}$	52.80	80.66
	$\mathcal{SS}$	52.80	**86.15**

Source domain was synthetic and target domain was real images with 10 images per-class used for training. The metrics for both G2A and SE were reported from the original source. The results for CCSA were obtained from our own implementation

3.3.3 VisDA-C Dataset

VisDA-C [19] dataset contains a synthetic domain and a real domain. Only a few numbers has been reported for this new datasets in literature, including G2A [23], SE [4], and CCSA [16]. For experiments, we followed the evaluation protocol in G2A [23]. Classification results are shown as Table 3.5. Under a supervised setting, i.e., 10 samples per class from the target domain were used for training, *d*-SNE achieved higher accuracy than CCSA and G2A, which is an unsupervised method that used all unlabeled data in the target domain. Under the same semi-supervised setting, *d*-SNE outperformed SE [4]. Feature visualizations in Fig. 3.4 indicates that *d*-SNE not only can align features across domains but making them discriminative.

3.4 Conclusion

With the advent of large quantities of data, domain adaptation has recently gained great success, especially for deep learning-based approaches. In this chapter, we introduced a novel supervised domain adaptation based on the stochastic neighborhood embedding. With the ability to align class to class between the source and target domains, the proposed method is robust to large domain shift even when the training data is extremely limited. The semi-supervised extension allows us to further improve the model performance by incorporating unlabeled data for training. Extensive experiments demonstrate that out method achieved state-of-the-art results in most of the benchmark datasets.

Fig. 3.4 t-SNE visualization of *d*-SNE's latent-embedding space for the VisDA-C dataset. (**a**) Embeddings produced by the model trained with source images only. (**b**) Embeddings produced by the model trained with target images only and (**c**) The joint latent-embedding space of *d*-SNE. Different colors represent different classes. Embeddings from the source and target domains are indicated by circles and stars, respectively

References

1. Bousmalis, K., Silberman, N., Dohan, D., Erhan, D., Krishnan, D.: Unsupervised pixel-level domain adaptation with generative adversarial networks. In: Proceedings of the IEEE Conference on Computer Vision and Pattern Recognition, pp. 3722–3731 (2017)
2. Deng, J., Dong, W., Socher, R., Li, L.J., Li, K., Fei-Fei, L.: ImageNet: a large-scale hierarchical image database. In: Proceedings of Computer Vision and Pattern Recognition (2009)
3. Ding, Z., Li, S., Shao, M., Fu, Y.: Graph adaptive knowledge transfer for unsupervised domain adaptation. In: Proceedings of the European Conference on Computer Vision (ECCV), pp. 37–52 (2018)
4. French, G., Mackiewicz, M., Fisher, M.: Self-ensembling for visual domain adaptation. In: Proceedings of International Conference on Learning Representations, Vancouver, pp. 1–15 (2018). https://openreview.net/pdf?id=rkpoTaxA-
5. Ganin, Y., Ustinova, E., Ajakan, H., Germain, P., Larochelle, H., Laviolette, F., Marchand, M., Lempitsky, V.: Domain-adversarial training of neural networks. J. Mach. Learn. Res. **17**(1), 2096–2030 (2016)
6. Ghifary, M., Kleijn, W.B., Zhang, M., Balduzzi, D., Li, W.: Deep reconstruction-classification networks for unsupervised domain adaptation. In: European Conference on Computer Vision, pp. 597–613. Springer, Berlin (2016)
7. Goodfellow, I., Pouget-Abadie, J., Mirza, M., Xu, B., Warde-Farley, D., Ozair, S., Courville, A., Bengio, Y.: Generative adversarial nets. In: Advances in Neural Information Processing Systems, pp. 2672–2680 (2014)
8. Haeusser, P., Frerix, T., Mordvintsev, A., Cremers, D.: Associative domain adaptation. In: Proceedings of the IEEE International Conference on Computer Vision, pp. 2765–2773 (2017)
9. He, K., Zhang, X., Ren, S., Sun, J.: Identity mappings in deep residual networks. In: European Conference on Computer Vision, pp. 630–645. Springer, Berlin (2016)
10. Hinton, G.E., Roweis, S.T.: Stochastic neighbor embedding. In: Advances in Neural Information Processing Systems, pp. 857–864 (2003)
11. Hoffman, J., Tzeng, E., Park, T., Zhu, J.Y., Isola, P., Saenko, K., Efros, A., Darrell, T.: Cycada: cycle-consistent adversarial domain adaptation. In: Proceedings of the 35th International Conference on Machine Learning (2018)
12. Hull, J.J.: A database for handwritten text recognition research. IEEE Trans. Pattern Anal. Mach. Intell. **16**(5), 550–554 (1994)
13. LeCun, Y., Bottou, L., Bengio, Y., Haffner, P., et al.: Gradient-based learning applied to document recognition. Proc. IEEE **86**(11), 2278–2324 (1998)
14. Liu, M.Y., Tuzel, O.: Coupled generative adversarial networks. In: Proceedings of Advances in Neural Information Processing Systems, pp. 469–477 (2016). arXiv:1606.07536. http://arxiv.org/abs/1606.07536
15. Motiian, S., Jones, Q., Iranmanesh, S., Doretto, G.: Few-shot adversarial domain adaptation. In: Advances in Neural Information Processing Systems, pp. 6670–6680 (2017)
16. Motiian, S., Piccirilli, M., Adjeroh, D.A., Doretto, G.: Unified deep supervised domain adaptation and generalization. In: Proceedings of the IEEE International Conference on Computer Vision, pp. 5715–5725 (2017)
17. Murez, Z., Kolouri, S., Kriegman, D., Ramamoorthi, R., Kim, K.: Image to image translation for domain adaptation. In: Proceedings of the IEEE Conference on Computer Vision and Pattern Recognition, pp. 4500–4509 (2018)
18. Netzer, Y., Wang, T.: Reading digits in natural images with unsupervised feature learning. In: Proceedings of Advances in Neural Information Processing Systems, Granada, pp. 1–9 (2011)
19. Peng, X., Usman, B., Kaushik, N., Wang, D., Hoffman, J., Saenko, K.: Visda: a synthetic-to-real benchmark for visual domain adaptation. In: Proceedings of the IEEE Conference on Computer Vision and Pattern Recognition Workshops, pp. 2021–2026 (2018)
20. Rozantsev, A., Salzmann, M., Fua, P.: Beyond sharing weights for deep domain adaptation. IEEE Trans. Pattern Anal. Mach. Intell. **41**(4), 801–814 (2018)

21. Russo, P., Carlucci, F.M., Tommasi, T., Caputo, B.: From source to target and back: symmetric bi-directional adaptive GAN. In: Proceedings of the IEEE Conference on Computer Vision and Pattern Recognition, pp. 8099–8108 (2018)
22. Saenko, K., Kulis, B., Fritz, M., Darrell, T.: Adapting visual category models to new domains. In: Proceedings of European Conference on Computer Vision, Grete, pp. 213–226 (2010)
23. Sankaranarayanan, S., Balaji, Y., Castillo, C.D., Chellappa, R.: Generate to adapt: aligning domains using generative adversarial networks. In: Proceedings of the IEEE Conference on Computer Vision and Pattern Recognition, pp. 8503–8512 (2018)
24. Sener, O., Song, H.O., Saxena, A., Savarese, S.: Learning transferrable representations for unsupervised domain adaptation. In: Advances in Neural Information Processing Systems, pp. 2110–2118 (2016)
25. Shu, R., Bui, H.H., Narui, H., Ermon, S.: A DIRT-T approach to unsupervised domain adaption. In: Proceedings of International Conference on Learning Representations, pp. 1–19 (2018)
26. Simonyan, K., Zisserman, A.: Very deep convolutional networks for large-scale image recognition (2014). Preprint. arXiv:1409.1556
27. Tarvainen, A., Valpola, H.: Mean teachers are better role models: weight-averaged consistency targets improve semi-supervised deep learning results. In: Proceedings of Advances in Neural Information Processing Systems, Long Beach (2017). http://arxiv.org/abs/1703.01780
28. Tzeng, E., Hoffman, J., Darrell, T., Saenko, K., Lowell, U.: Simultaneous deep transfer across domains and tasks. In: Proceedings of International Conference on Computer Vision, Las Condes (2015). https://doi.org/10.1109/ICCV.2015.463. https://www.robots.ox.ac.uk/~vgg/rg/papers/Tzeng_ICCV2015.pdf
29. Tzeng, E., Hoffman, J., Saenko, K., Darrell, T.: Adversarial discriminative domain adaptation. In: Proceedings of IEEE Conference on Computer Vision and Pattern Recognition, Honolulu, pp. 2962–2971 (2017). https://doi.org/10.1109/CVPR.2017.316
30. Weinberger, K.Q., Saul, L.K.: Distance metric learning for large margin nearest neighbor classification. J. Mach. Learn. Res. **10**(Feb), 207–244 (2009)
31. Wen, Y., Zhang, K., Li, Z., Qiao, Y.: A discriminative feature learning approach for deep face recognition. In: European Conference on Computer Vision, pp. 499–515. Springer, Berlin (2016)
32. Xu, X., Zhou, X., Venkatesan, R., Swaminathan, G., Majumder, O.: d-SNE: domain adaptation using stochastic neighborhood embedding. In: Proceedings of the IEEE Conference on Computer Vision and Pattern Recognition, pp. 2497–2506 (2019)
33. Zhou, X., Prasad, S.: Domain adaptation for robust classification of disparate hyperspectral images. IEEE Trans. Comput. Imaging **3**(4), 822–836 (2017)
34. Zhu, J.Y., Park, T., Isola, P., Efros, A.A.: Unpaired image-to-image translation using cycle-consistent adversarial networks. In: Proceedings of the IEEE International Conference on Computer Vision, pp. 2223–2232 (2017)

Chapter 4
Deep Hashing Network for Unsupervised Domain Adaptation

**Raghavendran Ramakrishnan, Bhadrinath Nagabandi, Jose Eusebio,
Shayok Chakraborty, Hemanth Venkateswara,
and Sethuraman Panchanathan**

4.1 Introduction

Recent advances in deep learning have enabled us to develop solutions for various computer vision applications that need not be programmed explicitly. However, these impressive performances are achieved only on the tasks with vast amounts of labeled data. Moreover, collecting and annotating sufficient data for every task is prohibitively expensive and nearly impossible in many cases. For example, consider an application in the medical domain, where labeling the data requires domain expertise. It becomes costly and time-consuming to hire domain experts to label large amounts of data. However, in many cases, labeled data from a different but relevant domain is available. It is essential to leverage such labeled data from a relevant domain to reduce the demand for labeled samples in the target domain.

Raghavendran Ramakrishnan and Bhadrinath Nagabandi equally contributed to this work.

R. Ramakrishnan · B. Nagabandi · H. Venkateswara (✉) · S. Panchanathan
Arizona State University, Tempe, AZ, USA
e-mail: rramak11@asu.edu; bnagaban@asu.edu; hemanthv@asu.edu; panch@asu.edu

J. Eusebio
Axosoft, Scottsdale, AZ, USA
e-mail: miggye@axosoft.com

S. Chakraborty
Florida State University, Tallahassee, FL, USA
e-mail: shayok@cs.fsu.edu

© Springer Nature Switzerland AG 2020
H. Venkateswara, S. Panchanathan (eds.), *Domain Adaptation in Computer Vision with Deep Learning*, https://doi.org/10.1007/978-3-030-45529-3_4

Pre-trained deep neural networks undergo performance degradation when applied to a target domain that has dissimilar statistics from the source domain on which the model was initially trained [43]. This drop in performance is a consequence of the domain shift that exists between the source and the target domains. The domain shift often occurs due to several factors such as illumination, pose, resolution, background, etc. Domain adaptation algorithms overcome this domain shift and leverage labeled data from a source domain to develop robust models that can accurately discriminate the samples of the target domain. Furthermore, unsupervised domain adaptation is a more challenging paradigm where the source domain is fully labeled and the target domain is unlabeled.

The conventional non-deep learning based approaches in domain adaptation (shallow domain adaptation) follow a two-step process. The first step is to extract meaningful features from the data, and the second step involves training a model to align the features of the source and target domains. These are static procedures where the features are initially extracted and then aligned. With the advent of deep learning, the domain adaptation algorithms have exploited the deep network's capability to learn transferable feature representations. As the features learned in these deep neural networks are data and task-specific, they demonstrate a significant improvement in the empirical performance over conventional shallow methods.

With the exponential growth of data available on the Internet, demand for efficient data storage and retrieval algorithms has also increased. One of the techniques that is studied widely for these purposes is hashing. Hashing provides the most flexible method of data retrieval due to its fast query speed and low memory overhead. Hashing techniques convert a very high dimensional data to a low dimensional compact binary code. Moreover, these techniques generate a similar binary encoding for similar items. In this chapter we learn hash-based image feature representations using a deep neural network. These representations allow for efficient and robust similarity measures between data points.

This chapter discusses a Domain Adaptive Hashing framework (DAH) that generates a binary hash code instead of a probability distribution over all the classes for each data point in the input space. We see multiple advantages in estimating a hash code instead of a standard probability distribution. (1) These hash codes are used to develop a unique loss function for the target data in the absence of the labels. (2) Label predictions are found to be more robust with hash codes as compared to probability vectors. In this framework, a deep neural network is trained end-to-end by leveraging the labeled data from the source domain and the unlabeled data from the target domain. During inference, as the model outputs the hash code instead of a probability vector, we compare the hash code of the given test samples with the hash codes of the training samples to arrive at a robust category prediction. The DAH framework proposes a unique loss function that includes multiple components: (1) a supervised hash loss using the labeled data of the source domain, (2) an unsupervised entropy loss for the unlabeled data of the target domain, and (3) a multi-kernel Maximum Mean Discrepancy (MK-MMD) loss to align the source and target distributions. This chapter is partially based on earlier work from [42], where we also introduced the *Office-Home* dataset which has become very popular in recent years for evaluating the performance of domain adaptation algorithms.

4.2 Related Work

One of the critical challenges in domain adaptation is the problem of domain shift that exists between the source and target domains. In the past, several approaches have been explored to address this problem. They can broadly be classified into shallow and deep approaches. In the shallow approach (non-deep learning), the features used are predetermined, and the classifier trained on the source data is adapted to the target data [1, 4]. Other approaches use a transformation matrix to linearly transform the source data so that it is well-aligned with the target data [27, 38]. Some other works like [9, 10, 21], weight the source samples and assign a higher value to data points that are similar to the target samples, leading to better alignment of the domains.

A common approach in domain adaptation is to project the source and target data to a common subspace and align their principal axes to reduce the domain discrepancy [16, 39]. Discrepancy estimation metrics such as Maximum Mean Discrepancy (MMD) provide a distribution divergence between the two domains. This shift is measured by projecting the data sets to a reproducing kernel Hilbert space [13]. Nonlinear domain alignment with MMD is achieved using Kernel PCA methods [30, 31, 36]. Another common approach in computer vision is to learn a transformation between the source and target domain by treating a domain subspace as a point on a manifold [20, 23]. A detailed survey of additional approaches is found in [35, 37].

Unlike the shallow techniques where the features are pre-determined, deep convolutional neural networks (CNNs) have been very successful in learning data-specific discriminative features [8]. The ability to capture and disentangle factors of variation makes CNNs an excellent tool for multi-task transfer learning [2, 12, 18, 34]. The lower layers of a CNN capture generic features and the upper layers are responsible for task-specific features, [43]. A few approaches have applied this understanding to align the features from the fully-connected deep layers of a neural network to achieve domain adaptation, [32, 40]. With the advent of generative adversarial networks (GANs), Ganin et al. [17] proposed Domain Adversarial Networks with a gradient reversal layer to learn domain invariant features. In recent years adversarial feature alignment has proved to be the most popular approach for domain adaptation.

Unsupervised hashing techniques are proposed to generate unique hash codes for efficient storage and retrieval of data [22, 25]. Earlier approaches that used neural networks to generate hash codes have produced state-of-the-art results [7, 11, 14]. A related work using hashing and adaptation appears in cross-modal hashing, where deep hashing techniques embed multi-modal data and learn hash codes for two related domains, like text and images [5, 6, 28]. Deep Adaptive Hashing (DAH) network is one of the first approaches to use deep hashing networks for unsupervised domain adaptation.

4.3 Domain Adaptive Hashing Network

In standard unsupervised domain adaptation we assume the data is from two domains: the source and the target. Let $\mathcal{D}_s = \{x_i^s, y_i^s\}_{i=1}^{n_s}$ be the source domain with n_s labeled examples and $\mathcal{D}_t = \{x_i^t, y_i^t\}_{i=1}^{n_t}$ be the target domain with n_t labeled examples. In this chapter, we deal with unsupervised domain adaptation where the labels y_i^t in the target domain are unknown. The data points x_i^* belong to X, where X is some input space. For a given x_i^* the corresponding label is represented by $y_i^* \in Y := \{1, \ldots, C\}$. The goal of unsupervised domain adaptation is to overcome the domain shift between the source and target domains and learn a classifier that accurately classifies the unlabeled data in the target domain. In this work, we minimize domain shift using a distribution-distance metric and learn a robust mapping $\psi(.)$ using the labeled source data in order to correctly classify the images of the target domain.

We use a pre-trained neural network (VGG-F [8]) with 5 convolutional layers (*conv*1–*conv*5) and 3 fully connected layers (*fc*6–*fc*8) followed by a loss layer. Since our model has to predict hash codes instead of softmax probabilities, we replace the standard *fc*8 layer from the pre-trained network with a newly introduced *hash-fc*8 layer. The output of *hash-fc*8 layer is represented as $\psi(x)$, where $\psi(x) \in \mathbb{R}^d$, which we further convert to a hash code $h = \text{sgn}(\psi(x))$, where sgn(.) is the sign function. Thus, the output of the network is a binary hash code h_i, for every data point x_i, where $h_i \in \{-1, +1\}^d$. We summarize the overall architecture in Fig. 4.1. We train the model with a combination of multiple loss functions, (1) *MK-MMD loss* for domain alignment, (2) *supervised hash loss* for the labeled source data, and (3) *unsupervised entropy loss* for the unlabeled target data.

4.3.1 Reducing Domain Disparity

Deep layers in a neural network generate task-specific representations while the initial layers output generic transferable representations [43]. We apply this understanding to align the output representations of the source and the target in the fully connected layers $\mathcal{F} = \{fc6, fc7, fc8\}$. We implement the Multi-kernel Maximum Mean Discrepancy (MK-MMD) measure to reduce distribution discrepancy [32, 33]. The MK-MMD loss is given by,

$$\mathcal{M}(\mathcal{U}_s, \mathcal{U}_t) = \sum_{l \in \mathcal{F}} d_k^2(u_s^l, u_t^l), \tag{4.1}$$

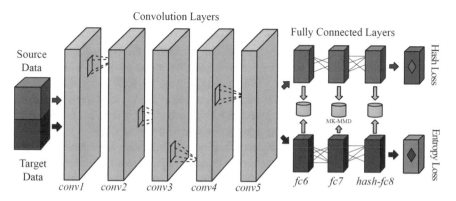

Fig. 4.1 The Domain Adaptive Hash (DAH) network that outputs hash codes for the source and the target. We train the network with a batch of source and target data. The convolution layers *conv1–conv5* and the fully connected layers *fc6* and *fc7* are fine-tuned from the VGG-F network. The MK-MMD loss trains the DAH to learn feature representations which align the source and the target. The *hash-fc8* layer is trained to output vectors of d dimensions. The supervised hash loss drives the DAH to estimate a unique hash value for each object category. The unsupervised entropy loss aligns the target hash values to their corresponding source categories. Best viewed in color (*Image Credit: [42]*)

The MK-MMD measure $d_k^2(.)$ [24] is computed over the output representations of the source and target data at layer l which is given by $\mathcal{U}_s^l = \{u_i^{s,l}\}_{i=1}^{n_s}$ and $\mathcal{U}_t^l = \{u_i^{t,l}\}_{i=1}^{n_t}$, where $u_i^{*,l}$ is the output representation of x_i^* for the lth layer. The final layer outputs are denoted as $\mathcal{U}_s$ and $\mathcal{U}_t$. For a nonlinear mapping $\phi(.)$ of reproducing kernel Hilbert space $\mathcal{H}_k$ and kernel $k(.)$, where $k(x, y) = \langle \phi(x), \phi(y) \rangle$, the MMD is given by,

$$d_k^2(\mathcal{U}_s^l, \mathcal{U}_t^l) = \left\| \mathbb{E}[\phi(u^{s,l})] - \mathbb{E}[\phi(u^{t,l})] \right\|_{\mathcal{H}_k}^2. \tag{4.2}$$

The kernel $k(.)$, is a convex combination of κ positive semi-definite kernels, $\{k_m\}_{m=1}^{\kappa}$, $\mathcal{K} := \left\{ k : k = \sum_{m=1}^{\kappa} \beta_m k_m, \sum_{m=1}^{\kappa} \beta_m = 1, \beta_m \geq 0, \forall m \right\}$. We set $\beta_m = 1/\kappa$ according to [33].

4.3.2 Supervised Hashing for Source

We devise a supervised hashing loss for the source data since it is labeled. Let us consider two source data points x_i and x_j and let h_i, h_j be their corresponding hash values. The hash values take one of two discrete values $h_i \in \{-1, +1\}^d$. To determine if any two given source data points belong to the same class, we compute the difference between their corresponding hash codes. The hamming distance can be used as a distance metric for computing the difference. The hamming distance

computes the number of dimensions that are dissimilar between the two vectors. It
is defined as $\text{dist}_H(\boldsymbol{h}_i, \boldsymbol{h}_j) = \frac{1}{2}(d - \boldsymbol{h}_i^\top \boldsymbol{h}_j)$, where d is the number of dimensions.
The dot product $\boldsymbol{h}_i^\top \boldsymbol{h}_j$ measures the similarity between the two vectors in terms
of d. Consider two input source samples belonging to the same class. Their hash
codes should be similar thereby yielding a high value for their dot product. It can
be inferred that $\text{dist}_H(.)$ decreases with the increase in the value of the dot product.
Conversely, when the distance between the two hash codes is small, there is a high
probability that the corresponding samples belong to the same class.

We define a $n_s \times n_s$ similarity matrix $S = \{s_{ij}\}$ over all the source samples.
Element s_{ij} in S is the similarity between source data point i and j with $s_{ij} \in \{0, 1\}$.
If both the samples belong to the same category then the value of $s_{ij} = 1$ and
0 otherwise. The probability of similarity between the samples $\boldsymbol{x}_i, \boldsymbol{x}_j$ with their
corresponding hash codes $\boldsymbol{h}_i, \boldsymbol{h}_j$ is defined as:

$$p(s_{ij}|\boldsymbol{h}_i, \boldsymbol{h}_j) = \begin{cases} \sigma(\boldsymbol{h}_i^\top \boldsymbol{h}_j), & s_{ij} = 1 \\ 1 - \sigma(\boldsymbol{h}_i^\top \boldsymbol{h}_j), & s_{ij} = 0, \end{cases} \tag{4.3}$$

where $\sigma(x) = \frac{1}{1+e^{-x}}$ is the sigmoid activation function. Let $\mathbf{H} = \{\boldsymbol{h}_i\}_{i=1}^{n_s}$ be the
set of hash codes for the source data points that the network outputs. With the
assumption that the samples are independent and identically distributed (iid), we
can minimize the negative log likelihood of S with respect to $\mathbf{H}$, so that the output
hash codes are consistent with the similarity between the data points;

$$\min_{\mathbf{H}} \mathcal{L}(\mathbf{H}) = -\log p(S|\mathbf{H})$$

$$= -\sum_{s_{ij} \in S} \left(s_{ij} \boldsymbol{h}_i^\top \boldsymbol{h}_j - \log(1 + \exp(\boldsymbol{h}_i^\top \boldsymbol{h}_j)) \right). \tag{4.4}$$

However, Eq. (4.4) is a discrete optimization problem that is hard to solve because
the hash codes $\boldsymbol{h}_i \in \{-1, 1\}^d$ take only discrete values. We relax the discrete
constraint on the hash codes and instead solve for $\boldsymbol{u}_i \in \mathbb{R}^d$, where $\mathcal{U}_s = \{\boldsymbol{u}_i\}_{i=1}^{n_s}$
is the output of the network with $\boldsymbol{u}_i = \psi(\boldsymbol{x}_i)$. We drop the superscript notation for
ease of representation.

We encounter the following challenges due to this relaxation. (1) Substituting
$\langle \boldsymbol{h}_i, \boldsymbol{h}_j \rangle$ with $\langle \boldsymbol{u}_i, \boldsymbol{u}_j \rangle$, introduces an approximation error. (2) When the real-valued
outputs $\boldsymbol{u}_i$ are binarized, we introduce quantization errors, [44]. To account for
the approximation error, we use tanh(.) as the activation of the final layer. This
bounds the components of $\boldsymbol{u}_i$ between -1 and $+1$. We address quantization error by
minimizing $\|\boldsymbol{u}_i - \text{sgn}(\boldsymbol{u}_i)\|_2^2$, [22]. More formally, the supervised hashing objective
can be written as,

$$\min_{\mathcal{U}_s} \mathcal{L}(\mathcal{U}_s) = -\sum_{s_{ij} \in S} \left(s_{ij} \boldsymbol{u}_i^\top \boldsymbol{u}_j - \log(1 + \exp(\boldsymbol{u}_i^\top \boldsymbol{u}_j)) \right)$$

$$+ \sum_{i=1}^{n_s} ||\boldsymbol{u}_i - \text{sgn}(\boldsymbol{u}_i)||_2^2. \tag{4.5}$$

4.3.3 Unsupervised Hashing for Target

We cannot apply the supervised hashing loss to the target data because there are no labels available during training. Hence, we rely only on the similarity measures $\langle \boldsymbol{u}_i, \boldsymbol{u}_j \rangle$ to guide the network to classify the target data. Without loss of generality we assume a batch of data consists of source data points with K examples per category for each of the C categories. We represent the source output data points which belong to a category j as $(\{\boldsymbol{u}_k^{s_j}\}_{k=1}^K)$. The target data point output $\boldsymbol{u}_i^t$ is said to belong to the jth category when it is similar to most of the source output data points $(\{\boldsymbol{u}_k^{s_j}\}_{k=1}^K)$. It must also be dissimilar to every other source output data point $\boldsymbol{u}_k^{s_l}$ that belongs to a different category ($j \neq l$). We therefore define the probability of the target data point $\boldsymbol{x}_i$ belonging to class j as,

$$p_{ij} = \frac{\sum_{k=1}^K \exp(\boldsymbol{u}_i^{t\top} \boldsymbol{u}_k^{s_j})}{\sum_{l=1}^C \sum_{k=1}^K \exp(\boldsymbol{u}_i^{t\top} \boldsymbol{u}_k^{s_l})} \tag{4.6}$$

The numerator in (4.6) measures the similarity of the target sample with K source samples from class j. The denominator computes the similarity of the target data point with K source samples from each of the C classes and is a normalizing factor to ensure $\sum_j p_{ij} = 1$. For every target output data point $\boldsymbol{u}_i^t$, we can define a probability vector $\boldsymbol{p}_i = [p_{i1}, \ldots, p_{iC}]^T$, where p_{ij} is the probability of $\boldsymbol{u}_i^t$ belonging to class j. In an ideal scenario, the probability vector resembles a one-hot vector with a value 1 assigned to the true class of the sample and a value 0 assigned to other classes.

A one hot vector can be viewed as a low entropy representation of $\boldsymbol{p}_i$, where we are confident about the category of the target data point. We introduce a loss term to minimize the entropy of these probability vectors thereby ensuring the network is confident with the target category assignment. The entropy loss for the network outputs is given by,

$$\mathcal{H}(\mathcal{U}_s, \mathcal{U}_t) = -\frac{1}{n_t} \sum_{i=1}^{n_t} \sum_{j=1}^C p_{ij} \log(p_{ij}) \tag{4.7}$$

By minimizing Eq. (4.7), the probability vectors $\boldsymbol{p}_i$ tend towards a one hot representation. Enforcing this with K samples per class in the source ensures that the target sample $\boldsymbol{u}_i^t$ is similar to multiple (K) source samples from a specific category.

4.3.4 Domain Adaptive Hashing

The Domain Adaptive Hash (DAH) network applies unsupervised domain adaptation between the source and target using hashing (DAH). The model is a deep convolution neural network that incorporates supervised hashing for the source and unsupervised hashing for the target. The objective of the DAH network can be given as,

$$\min_{\mathcal{U}} \mathcal{J} = \mathcal{L}(\mathcal{U}_s) + \gamma \mathcal{M}(\mathcal{U}_s, \mathcal{U}_t) + \eta \mathcal{H}(\mathcal{U}_s, \mathcal{U}_t), \qquad (4.8)$$

where $\mathcal{U} := \{\mathcal{U}_s \cup \mathcal{U}_t\}$ and (γ, η) are the hyper-parameters for the MK-MMD loss and entropy loss respectively. They control the importance of domain alignment and target entropy loss. We obtain hash values from the output of the network as $\mathbf{H} = \text{sgn}(\mathcal{U})$. The MK-MMD loss (4.1), is calculated between $\{\mathcal{U}_s^l, \mathcal{U}_t^l\}$ at every fully connected layer $\{fc6, fc7, fc8\}$. The model is trained using the standard back-propagation algorithm. The Network architecture and other details are discussed below.

Network Architecture Deep neural networks are incredibly data-hungry besides it is not possible to collect and annotate millions of images for every task. Therefore, we apply transfer learning by choosing the pre-trained VGG-F network [8] that was trained initially on ImageNet 2012, a vast labeled dataset of 1000 categories which has over a million images. Current domain adaptation models are based on the more efficient ResNet architecture [26], which is a much better feature extractor compared to the VGG-F model. We believe the DAH would yield much better results if it were implemented using ResNet as the base network. The pre-trained VGG-F model has 5 convolution layers (conv1–conv5) followed by 3 fully connected layers (fc6, fc7, fc8). Our goal is to predict binary hash codes for a given data sample, and it is different from what the actual VGG-F model does. So, we modify the architecture by introducing a new layer hash-fc8 in place of the original fc8 layer. The binary hash codes we require are $\{-1, +1\}$, therefore we use a tanh() activation function to limit the predictions to the same range. Although tanh() is a smooth and zero-centered activation function with its output ranging between -1 and $+1$, it has a notable disadvantage of vanishing gradients in a deep neural network. When activations become saturated, gradients tend to vanish (value 0) and the network does not learn. As a workaround for this problem, we introduce a batch normalization layer before the tanh layer to prevent it from saturating. Hence, the final hash-fc8 layer is defined as $hash\text{-}fc8 := \{fc8 \rightarrow batch\text{-}norm \rightarrow tanh()\}$. Adding hash-fc8 as the final layer prevents activation saturation and provides greater stability for training the model.

4.4 Experiments and Results

We conduct extensive experiments to evaluate the DAH model for domain adaptation and hashing. We evaluate our model for (1) unsupervised domain adaptation, (2) discriminative features and, (3) unsupervised domain adaptive hashing. The implementation details are available at https://github.com/hemanthdv/da-hash.

4.4.1 Datasets

Office [38] This is one of the standard benchmark datasets for evaluating domain adaptation in computer vision. The dataset consists of images of objects from a regular office environment. It has 3 domains; Amazon (**A**), Dslr (**D**) and Webcam (**W**). This dataset has around 4100 images with a large majority of the images (2816 images) coming from the Amazon domain. In our experiments we adopt the standard protocol of evaluating every pair of transfer tasks for this dataset [32, 33]. We have 6 transfer tasks considering all combinations of source and target pairs given the 3 domains. $\mathbf{A} \rightarrow \mathbf{D}$, $\mathbf{D} \rightarrow \mathbf{A}$, $\mathbf{A} \rightarrow \mathbf{W}$, $\mathbf{W} \rightarrow \mathbf{A}$, $\mathbf{D} \rightarrow \mathbf{W}$ and $\mathbf{W} \rightarrow \mathbf{D}$. $\mathbf{A} \rightarrow \mathbf{D}$ implies, **A** is the source and **D** is the target.

Office-Home In the original work on which this chapter is based [42], we introduced this new dataset for domain adaptation. The *Office-Home* dataset has 4 domains, and each domain has images from 65 categories. The categories are from everyday objects with a total of around 15,500 images. The domains include, Art: artful renditions of objects in the form of paintings, sketches, ornamentation, etc.; Clipart: clipart images; Product: high-resolution images of objects without a background—similar to the Amazon category in the *Office* dataset; Real-World: images of objects gathered using a regular camera. We consider 12 transfer tasks for the Art (**Ar**), Clipart (**Cl**), Product (**Pr**) and Real-World (**Rw**) domains for all combinations of source and target for the 4 domains. $\mathbf{Ar} \rightarrow \mathbf{Cl}$, $\mathbf{Cl} \rightarrow \mathbf{Ar}$, $\mathbf{Ar} \rightarrow \mathbf{Pr}$, $\mathbf{Pr} \rightarrow \mathbf{Ar}$, $\mathbf{Ar} \rightarrow \mathbf{Rw}$, $\mathbf{Rw} \rightarrow \mathbf{Ar}$, $\mathbf{Cl} \rightarrow \mathbf{Pr}$, $\mathbf{Pr} \rightarrow \mathbf{Cl}$, $\mathbf{Cl} \rightarrow \mathbf{Rw}$, $\mathbf{Rw} \rightarrow \mathbf{Cl}$, $\mathbf{Pr} \rightarrow \mathbf{Rw}$ and $\mathbf{Rw} \rightarrow \mathbf{Pr}$. The dataset is available for research purposes at https://hemanthdv.github.io/officehome-dataset/.

4.4.2 Implementation Details

The model uses the MatConvnet framework for implementation [41]. We use a pre-trained VGG-F and fine-tune the weights of $conv1$–$conv5$, $fc6$ and $fc7$. The learning rate is set at $\frac{1}{10}$th the learning-rate used for *hash-fc8* and is varied between 10^{-4} to 10^{-5}. We use gradient descent with momentum 0.9 and weight decay set at 5×10^{-4} over 300 epochs. We use $K = 5$ samples from each category. Since the office dataset contains 31 categories, the input batch size for source is

155 (5 samples $\times$ 31 categories) samples. We randomly sample 155 images for the target. Hence, the total number of samples in a batch is 310. Similarly, for *Office-Home* dataset, the batch size is 650 (5 samples $\times$ 65 categories). This is cause for concern as the batch size increases with increase in the number of categories. We experimented with hash code length $d = 64$. The results for other hash code lengths follow similar trends as $d = 64$. Due to the imbalance in the number of like and unlike pairs in S, we boosted the similarity values for like-pairs so that $S_{i,j} \in \{0, 10\}$. The performance of the network improves with the increase in similarity weight for the like-pairs. The parameter $\eta = 1$ is used for entropy loss. The parameter γ is obtained through cross-validation using a binary domain classifier. We select γ which gives largest error on a validation set. For the MK-MMD loss we follow the heuristics outlined in [24]. We use a Gaussian kernel with a bandwidth σ calculated as the median of the pairwise distances in the training data. For the multi-kernel, the bandwidth is varied as $\sigma_m \in [2^{-8}\sigma, 2^8\sigma]$ multiplied with a factor of 2. The target data points are classified using $f(x_i^t) = p(y|\boldsymbol{h}_i^t)$. The classifier assigns the target data point to the class with maximum probability based on the hash codes of the source and the target.

4.4.3   Unsupervised Domain Adaptation

In this section, we study the performance of our Deep Hashing Network. Furthermore, we also compare the performance of our model with the existing state-of-the-art methods. We evaluate both our model and the other competitive baselines in an unsupervised domain adaptation setting, where the labeled data is available only in the source domain, and the target domain is fully unlabeled.

As part of the evaluation, we compare the performance of our model with the following classical shallow domain adaptation methods,

1. Geodesic Flow Kernel (**GFK**) [20]
2. Transfer Component Analysis (**TCA**) [36]
3. Correlation Alignment (**CORAL**) [39]
4. Joint Distribution Adaptation (**JDA**) [30],

and with deep learning methods,

1. Deep Adaption Network (**DAN**) [32]
2. Domain Adversarial Neural Networks (**DANN**) [17]

For the shallow learning methods, deep features extracted from the $fc7$ layer of the VGG-F network which was pre-trained on the ImageNet 2012 dataset, were used. We also study the effect of entropy loss in our method, i.e., the change in performance of the model by aligning the target hash values with the source values with similar representations. To be consistent with our analysis, we denote **DAH-e** as the model trained without the entropy loss and **DAH** is the full model.

Table 4.1 Recognition accuracies (%) for domain adaptation experiments on the *Office* dataset

Expt.	A→D	A→W	D→A	D→W	W→A	W→D	Avg.
GFK	48.59	52.08	41.83	89.18	49.04	93.17	62.32
TCA	51.00	49.43	48.12	93.08	48.83	96.79	64.54
CORAL	54.42	51.70	48.26	95.97	47.27	98.59	66.04
JDA	59.24	58.62	51.35	96.86	52.34	97.79	69.37
DAN	67.04	67.80	50.36	95.85	52.33	99.40	72.13
DANN	72.89	72.70	56.25	96.48	53.20	99.40	75.15
DAH-e	66.27	66.16	55.97	94.59	53.91	96.99	72.31
DAH	66.47	68.30	55.54	96.10	53.02	98.80	73.04

A Amazon, *D* Dslr, *W* Webcam
A→W implies A is source and W is target (*Table Credit: [42]*)

Results and Discussion In Tables 4.1 and 4.2, we report the classification accuracies (%) of all the models, including ours on both the *Office* and *Office-Home* datasets. The values in the table indicate the number of samples correctly classified in percentages in the target domain. We compare our Deep Hashing Network with a hash length $d = 64$ bits. It can be observed that the DAH model consistently produces better results than other baselines in multiple transfer tasks from the *Office* dataset. However, for the *Office* dataset the DANN surpasses the DAH. We believe that domain adversarial training is more effective than DAH when the number of categories in the dataset are fewer. Since domain alignment is category agnostic, the aligned domains are possibly not classification friendly in the presence of a large number of categories. These results lead us to the conclusion that the DAH performs better when the number of categories in the dataset is large.

The DAH-e row shows the accuracy when the model is trained with η set to zero, i.e. without entropy. We note that DAH delivers better performance than DAH-e because minimizing the entropy on the target data using (4.7) helps in aligning the target samples with the source samples of similar representations. Therefore, the improved alignment of the source and target samples is responsible for the boost in the accuracy.

4.4.4 Feature Analysis

In Fig. 4.2 we show that the $\mathcal{A}$-distance between the source and target domains using VGG-F, DAN and DAH features. The $\mathcal{A}$-distance provides an estimate of the discrepancy between the two domains which is approximately equivalent to $2(1-2\epsilon)$, where ϵ is the generalization error of a binary classifier trained to classify between the source and target domains [3].

We use a LIBLINEAR SVM [15] as the classifier along with fivefold cross-validation to estimate the value of ϵ. The values of the bars in Fig. 4.2 indicate the

Table 4.2 Recognition accuracies (%) for domain adaptation experiments on the *Office-Home* dataset

Expt.	Ar→Cl	Ar→Pr	Ar→Rw	Cl→Ar	Cl→Pr	Cl→Rw	Pr→Ar	Pr→Cl	Pr→Rw	Rw→Ar	Rw→Cl	Rw→Pr	Avg.
GFK	21.60	31.72	38.83	21.63	34.94	34.20	24.52	25.73	42.92	32.88	28.96	50.89	32.40
TCA	19.93	32.08	35.71	19.00	31.36	31.74	21.92	23.64	42.12	30.74	27.15	48.68	30.34
CORAL	27.10	36.16	44.32	26.08	40.03	40.33	27.77	30.54	50.61	38.48	36.36	57.11	37.91
JDA	25.34	35.98	42.94	24.52	40.19	40.90	25.96	32.72	49.25	35.10	35.35	55.35	36.97
DAN	30.66	42.17	54.13	32.83	47.59	49.78	29.07	34.05	56.70	43.58	38.25	62.73	43.46
DANN	33.33	42.96	54.42	32.26	49.13	49.76	30.49	38.14	56.76	44.71	42.66	64.65	44.94
DAH-e	29.23	35.71	48.29	33.79	48.23	47.49	29.87	38.76	55.63	41.16	44.99	59.07	42.69
DAH	31.64	40.75	51.73	34.69	51.93	52.79	29.91	39.63	60.71	44.99	45.13	62.54	45.54

Ar Art, *Cl* Clipart, *Pr* Product, *Rw* Real-world
Ar→Cl implies Ar is source and Cl is target

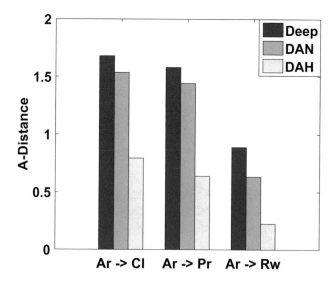

Fig. 4.2 $\mathcal{A}$-distances between source and target domains for Deep(VGG-F), DAN and DAH (*Image Credit: [42]*)

Fig. 4.3 t-SNE embeddings for 10 categories from Art (•) and Clipart(+) domains. Best viewed in color (*Image Credit: [42]*). (**a**) Deep features (Ar,Cl). (**b**) DAN features (Ar,Cl). (**c**) DAH features (Ar,Cl)

discrepancy between the domains. We can observe that in all the three transfer tasks, DAH produces better domain-aligned features with minimum distance between the domains.

The t-SNE embeddings of the features extracted from the penultimate layers of different models are presented in Fig. 4.3. We observe that DAH performs better at clustering the features into different classes compared to the other baselines. We can also observe the source and the target domain are also well overlapped as we move to DAN and DAH.

4.4.5 Unsupervised Domain Adaptive Hashing

In this section we evaluate the ability of the DAH to generate compact and efficient hash codes from the data for classifying unseen test instances in the absence of labels. There are promising unsupervised hashing techniques in literature, [7, 11, 19]. In a real-world setting, labels may become available through a different, but related (source) domain. In such cases, a procedure like the DAH that utilizes labeled data from a source domain to learn hash codes for the target domain is of immense practical importance. To address this real-world challenge we consider the following scenarios:

1. No labels are available and the hash codes are to be learned in an unsupervised manner. We compare with baseline unsupervised hashing methods (**ITQ**) [22] and (**KMeans**) [25] and state-of-the-art methods for unsupervised hashing (**BA**) [7] and (**BDNN**) [11].
2. Labeled data is made available from a related but different source domain. The labeled source data is used to learn hash codes for the target data. There is no domain adaptation and we refer to this method as **NoDA**. We deploy the deep pairwise-supervised hashing (DPSH) algorithm [29] trained with source data and apply the network to determine the hash values of the target.
3. We use labeled data from a related but different domain. We reduce the domain discrepancy between the domains as we learn the hash codes for the target using the **DAH**.
4. Labeled data is available in the target domain and we learn the hash values in a supervised manner. We refer to this procedure as **SuH**. This is a supervised technique and we use it merely to compare the relative performance of other approaches. We used the deep pairwise-supervised hashing (DPSH) algorithm [29] to train a deep network with the target data and evaluate the hash values of a validation set.

Results and Discussion Figures 4.4 and 4.5 show the precision-recall curves on each domain from the *Office-Home* dataset. For every domain in the *Office-Home* dataset, we train models in all four settings and visualize them in the same figure for a fair comparison. We also calculate the mean Average Precision (mAP) to evaluate the performance of the hashing method and report the values in Table 4.3 as per previous evaluation guidelines [7, 11, 19]. The numbers reported in Table 4.3, are estimated with a hash code of length $d = 64$ bits. When we used hash code lengths of $d = 16$ bits and $d = 128$ bits, we observed the same trend in the performance as reported in our experiments for $d = 64$ bits. The results observed from our experiments are very intuitive for all the models trained in the above settings. Because of the domain shift between the source and target domain, a model trained in **NoDA** setting had the most mediocre performance. It clearly shows that there is a need to learn a model that overcomes the domain shift and learns discriminative features. The methods **ITQ, KMeans, BA**, and **BDNN** trained in the unsupervised setting had slightly better performance than **NoDA**. This also highlights the role

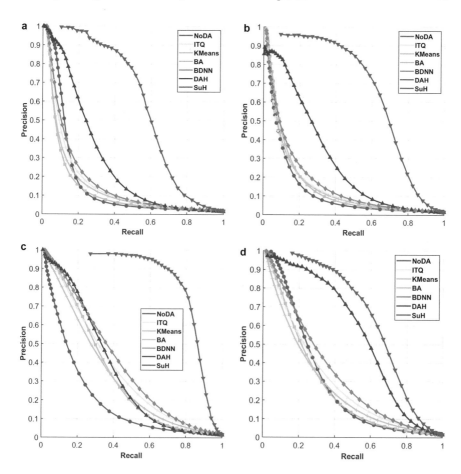

Fig. 4.4 Precision-Recall curves @64 bits for the *Office-Home* dataset. Comparison of hashing without domain adaptation (**NoDA**), shallow unsupervised hashing (**ITQ**, **KMeans**), state-of-the-art deep unsupervised hashing (**BA**, **BDNN**), unsupervised domain adaptive hashing (**DAH**) and supervised hashing (**SuH**). Best viewed in color (*Image Credit: [42]*). (**a**) Art. (**b**) Clipart. (**c**) Product. (**d**) Real-world

of negative-transfer in the **NoDA** case in the absence of domain alignment. The proposed **DAH** algorithm integrates both hash code learning for class discriminative features and domain adaptation for domain invariant features in a single framework. Hence, the learned features are more aligned between the domains and resulted in performance improvement on the target domain as shown in Figs. 4.4 and 4.5 and Table 4.3. Since the target labels are available for the **SuH** model, it performs the best as one can expect. Moreover, we note that the curves of **DAH** and the **SuH** are relatively similar which confirms the ability of the DAH model to generate meaningful hash codes and labels in the target domain.

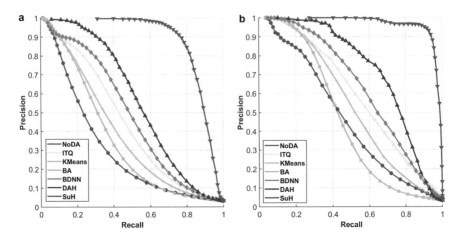

Fig. 4.5 Precision-Recall curves @64 bits for the *Office* dataset. Comparison of hashing without domain adaptation (**NoDA**), shallow unsupervised hashing (**ITQ, KMeans**), state-of-the-art deep unsupervised hashing (**BA, BDNN**), unsupervised domain adaptive hashing (**DAH**) and supervised hashing (**SuH**). Best viewed in color (*Image Credit: [42]*). (**a**) Amazon. (**b**) Webcam

Table 4.3 Mean average precision @64 bits

Expt.	NoDA	ITQ	KMeans	BA	BDNN	DAH	SuH
Amazon	0.324	0.465	0.403	0.367	0.491	0.582	0.830
Webcam	0.511	0.652	0.558	0.480	0.656	0.717	0.939
Art	0.155	0.191	0.170	0.156	0.193	0.302	0.492
Clipart	0.160	0.195	0.178	0.179	0.206	0.333	0.622
Product	0.239	0.393	0.341	0.349	0.407	0.414	0.774
Real-World	0.281	0.323	0.279	0.273	0.336	0.533	0.586
Avg.	0.278	0.370	0.322	0.301	0.382	0.480	0.707

For the NoDA and DAH results, Art is the source domain for Clipart, Product and Real-World and Clipart is the source domain for Art. Similarly, Amazon and Webcam are source target pairs (*Table Credit: [42]*)

4.5 Conclusions

In this paper, we proposed a Deep Hashing network, a model for unsupervised domain adaptation that exploits the capabilities of a deep neural network to produce domain invariant features and also learn efficient class discriminative hash codes. Our model solves 2 problems which are very common in the real world setting: Category assignment with less labeled data (with domain adaptation) and estimation of hash codes in an unsupervised setting (hash codes for target data). This work also introduced a new dataset *Office-Home* with more domains and categories for better evaluation of domain adaptation algorithms. This work is one of the first attempts to integrate hash code learning with unsupervised domain adaptation.

References

1. Aytar, Y., Zisserman, A.: Tabula rasa: model transfer for object category detection. In: IEEE ICCV (2011)
2. Bengio, Y., Courville, A., Vincent, P.: Representation learning: a review and new perspectives. IEEE Trans. Pattern Anal. Mach. Intell. **35**(8), 1798–1828 (2013)
3. Ben-David, S., Blitzer, J., Crammer, K., Kulesza, A., Pereira, F., Vaughan, J.W.: A theory of learning from different domains. Mach. Learn. **79**(1–2), 151–175 (2010)
4. Bruzzone, L., Marconcini, M.: Domain adaptation problems: a DASVM classification technique and a circular validation strategy. IEEE Trans. Pattern Anal. Mach. Intell. **32**(5), 770–787 (2010)
5. Cao, Y., Long, M., Wang, J., Yang, Q., Yu, P.S.: Deep visual-semantic hashing for cross-modal retrieval. In: ACM-SIGKDD (2016)
6. Cao, Z., Long, M., Yang, Q.: Transitive hashing network for heterogeneous multimedia retrieval. In: AAAI (2016)
7. Carreira-Perpinán, M.A., Raziperchikolaei, R.: Hashing with binary autoencoders. In: Proceedings of the IEEE Conference on Computer Vision and Pattern Recognition, pp. 557–566 (2015)
8. Chatfield, K., Simonyan, K., Vedaldi, A., Zisserman, A.: Return of the devil in the details: delving deep into convolutional nets. In: BMVC (2014)
9. Chattopadhyay, R., Sun, Q., Fan, W., Davidson, I., Panchanathan, S., Ye, J.: Multisource domain adaptation and its application to early detection of fatigue. ACM Trans. Knowl. Discov. Data **6**(4), 18 (2012)
10. Chu, W.S., De la Torre, F., Cohn, J.F.: Selective transfer machine for personalized facial action unit detection. In: Proceedings of the IEEE Conference on Computer Vision and Pattern Recognition, pp. 3515–3522 (2013)
11. Do, T.T., Doan, A.D., Cheung, N.M.: Learning to hash with binary deep neural network. In: European Conference on Computer Vision, pp. 219–234. Springer, Berlin (2016)
12. Donahue, J., Jia, Y., Vinyals, O., Hoffman, J., Zhang, N., Tzeng, E., Darrell, T.: Decaf: a deep convolutional activation feature for generic visual recognition. In: ICML, pp. 647–655 (2014)
13. Duan, L., Tsang, I.W., Xu, D.: Domain transfer multiple kernel learning. IEEE Trans. Pattern Anal. Mach. Intell. **34**(3), 465–479 (2012)
14. Erin Liong, V., Lu, J., Wang, G., Moulin, P., Zhou, J.: Deep hashing for compact binary codes learning. In: Proceedings of the IEEE Conference on Computer Vision and Pattern Recognition, pp. 2475–2483 (2015)
15. Fan, R.E., Chang, K.W., Hsieh, C.J., Wang, X.R., Lin, C.J.: Liblinear: a library for large linear classification. J. Mach. Learn. Res. **9**(Aug), 1871–1874 (2008)
16. Fernando, B., Habrard, A., Sebban, M., Tuytelaars, T.: Unsupervised visual domain adaptation using subspace alignment. In: CVPR, pp. 2960–2967 (2013)
17. Ganin, Y., Ustinova, E., Ajakan, H., Germain, P., Larochelle, H., Laviolette, F., Marchand, M., Lempitsky, V.: Domain-adversarial training of neural networks. J. Mach. Learn. Res. **17**(59), 1–35 (2016)
18. Glorot, X., Bordes, A., Bengio, Y.: Domain adaptation for large-scale sentiment classification: a deep learning approach. In: Proceedings of the 28th International Conference on Machine Learning (ICML-11), pp. 513–520 (2011)
19. Gong, Y., Lazebnik, S.: Iterative quantization: a procrustean approach to learning binary codes. In: 2011 IEEE Conference on Computer Vision and Pattern Recognition (CVPR), pp. 817–824. IEEE, Piscataway (2011)
20. Gong, B., Shi, Y., Sha, F., Grauman, K.: Geodesic flow kernel for unsupervised domain adaptation. In: IEEE CVPR (2012)
21. Gong, B., Grauman, K., Sha, F.: Connecting the dots with landmarks: discriminatively learning domain-invariant features for unsupervised domain adaptation. In: ICML (1), pp. 222–230 (2013)

22. Gong, Y., Lazebnik, S., Gordo, A., Perronnin, F.: Iterative quantization: a procrustean approach to learning binary codes for large-scale image retrieval. IEEE Trans. Pattern Anal. Mach. Intell. **35**(12), 2916–2929 (2013)
23. Gopalan, R., Li, R., Chellappa, R.: Domain adaptation for object recognition: an unsupervised approach. In: 2011 International Conference on Computer Vision, pp. 999–1006. IEEE, Piscataway (2011)
24. Gretton, A., Sejdinovic, D., Strathmann, H., Balakrishnan, S., Pontil, M., Fukumizu, K., Sriperumbudur, B.K.: Optimal kernel choice for large-scale two-sample tests. In: NIPS, pp. 1205–1213 (2012)
25. He, K., Wen, F., Sun, J.: K-means hashing: an affinity-preserving quantization method for learning binary compact codes. In: Proceedings of the IEEE Conference on Computer Vision and Pattern Recognition, pp. 2938–2945 (2013)
26. He, K., Zhang, X., Ren, S., Sun, J.: Deep residual learning for image recognition. In: Proceedings of the IEEE Conference on Computer Vision and Pattern Recognition, pp. 770–778 (2016)
27. Hoffman, J., Rodner, E., Donahue, J., Saenko, K., Darrell, T.: Efficient learning of domain-invariant image representations. In: ICLR (2013)
28. Jiang, Q.Y., Li, W.J.: Deep cross-modal hashing (2016). Preprint. arXiv:1602.02255
29. Li, W.J., Wang, S., Kang, W.C.: Feature learning based deep supervised hashing with pairwise labels. In: IJCAI (2016)
30. Long, M., Wang, J., Ding, G., Sun, J., Yu, P.S.: Transfer feature learning with joint distribution adaptation. In: Proceedings of the IEEE International Conference on Computer Vision, pp. 2200–2207 (2013)
31. Long, M., Wang, J., Ding, G., Sun, J., Yu, P.: Transfer joint matching for unsupervised domain adaptation. In: CVPR, pp. 1410–1417 (2014)
32. Long, M., Cao, Y., Wang, J., Jordan, M.: Learning transferable features with deep adaptation networks. In: ICML, pp. 97–105 (2015)
33. Long, M., Zhu, H., Wang, J., Jordan, M.I.: Unsupervised domain adaptation with residual transfer networks. In: NIPS (2016)
34. Oquab, M., Bottou, L., Laptev, I., Sivic, J.: Learning and transferring mid-level image representations using convolutional neural networks. In: Proceedings of the IEEE Conference on Computer Vision and Pattern Recognition, pp. 1717–1724 (2014)
35. Pan, S.J., Yang, Q.: A survey on transfer learning. IEEE Trans. Knowl. Data Eng. **22**(10), 1345–1359 (2010)
36. Pan, S.J., Tsang, I.W., Kwok, J.T., Yang, Q.: Domain adaptation via transfer component analysis. IEEE Trans. Neural Netw. **22**(2), 199–210 (2011)
37. Patel, V.M., Gopalan, R., Li, R., Chellappa, R.: Visual domain adaptation: a survey of recent advances. IEEE Signal Process. Mag. **32**(3), 53–69 (2015)
38. Saenko, K., Kulis, B., Fritz, M., Darrell, T.: Adapting visual category models to new domains. In: ECCV (2010)
39. Sun, B., Feng, J., Saenko, K.: Return of frustratingly easy domain adaptation. In: ICCV, TASK-CV (2015)
40. Tzeng, E., Hoffman, J., Darrell, T., Saenko, K.: Simultaneous deep transfer across domains and tasks. In: Proceedings of the IEEE International Conference on Computer Vision, pp. 4068–4076 (2015)
41. Vedaldi, A., Lenc, K.: Matconvnet – convolutional neural networks for matlab. In: Proceeding of the ACM International Conference on Multimedia (2015)
42. Venkateswara, H., Eusebio, J., Chakraborty, S., Panchanathan, S.: Deep hashing network for unsupervised domain adaptation. In: Proceedings of the IEEE Conference on Computer Vision and Pattern Recognition, pp. 5018–5027 (2017)
43. Yosinski, J., Clune, J., Bengio, Y., Lipson, H.: How transferable are features in deep neural networks? In: Advances in Neural Information Processing Systems, pp. 3320–3328 (2014)
44. Zhu, H., Long, M., Wang, J., Cao, Y.: Deep hashing network for efficient similarity retrieval. In: Thirtieth AAAI Conference on Artificial Intelligence (2016)

Chapter 5
Re-weighted Adversarial Adaptation Network for Unsupervised Domain Adaptation

Qingchao Chen, Yang Liu, Zhaowen Wang, Ian Wassell, and Kevin Chetty

5.1 Introduction

Deep networks have yielded state-of-the-art results from supervised learning applications in computer vision [13, 23, 41], however, they require a large amount of well-annotated training data, which is not always feasible to perform manually. Even with the help of annotations, trained models have difficulty to generalize well to new datasets caused by different factors of variation. The Domain Adaptation (DA) problem [36] was proposed in this context where the data distribution between the target domain (where a few of labels are available) and the source domain (well-annotated labels) varies so that the discriminative features and the classifiers in the source domain cannot be transferred to the target domain [36, 48]. Unsupervised Domain Adaptation (UDA) methods aim to address the dataset bias in real-world applications where the data distribution between the target domain (without label information) and the source domain (well-annotated labels) need to be matched so that the classifiers trained from the source domain can be transferred to the target

Q. Chen (✉) · Y. Liu
University of Oxford, Oxford, UK
e-mail: qingchao.chen@eng.ox.ac.uk; yang.liu@eng.ox.ac.uk

Z. Wang
Adobe Research, San Jose, CA, USA
e-mail: zhawang@adobe.com

I. Wassell
University of Cambridge, Cambridge, UK
e-mail: ijw24@cam.ac.uk

K. Chetty
University College London, London, UK
e-mail: k.chetty@ucl.ac.uk

© Springer Nature Switzerland AG 2020
H. Venkateswara, S. Panchanathan (eds.), *Domain Adaptation in Computer Vision with Deep Learning*, https://doi.org/10.1007/978-3-030-45529-3_5

domain. Successful methods firstly match the feature distribution and secondly adapt the classifier from source to target domains.

Since deep features have been shown to be more transferable compared with shallow features among datasets, a line of research has investigated measurements to estimate marginal distribution divergence of deep features, e.g., the un-biased estimate of marginal distribution divergence, Maximum Mean Discrepancy (MMD) [19, 30–32, 49]; currently the best-performing adversarial training methods [6, 16, 26, 45] and the more generalized f-divergence [35]. Arguably, both the MMD and f-divergence based methods are based on the assumption that feature distributions of the source and target domain at least should share a common support. This is an unrealistic condition since this can rarely be met in the real-world adaptation tasks due to unseen and complex factors that cause the domain discrepancies [10], such as light conditions, acquisition devices or even from different image formats e.g., RGB and HHA. From this point of view, MMD and currently f-divergence based methods fail to adapt between domains once their distributions do not have significant overlap. To alleviate the need for a common support and to develop the more generalized adaptation methods, optimal transport (OT) based methods have been proposed to match the source and target feature distributions by minimizing the global transportation efforts [10, 11]. However, these methods have not been formalized and embedded into an end-to-end pipeline, which limits its applicability to large-scale UDA problems.

Besides selecting a good divergence measure of the marginal feature distribution, another line of research focused on integrating the class label prediction with the marginal features, enabling classifier adaptation. Long et al. [29, 32] and Courty et al. both [10] proposed to match the joint distribution of feature and label by embedding the label prediction and the marginal features via either learning the joint cross-domain metrics or advanced feature fusion methods.

In this paper, we propose a Re-weighted Adversarial Adaptation Network (RAAN) for UDA problems including the following two main contributions:

1. To match marginal feature distributions across disparate domains, we train a domain discriminator network together with the deep feature extractors in an adversarial manner to minimize the OT based Earth Mover (EM) distance. Compared with other methods adopting geometry-oblivious measures, RAAN can better reduce large feature distribution divergence.
2. For classifier adaptation and integrating the discriminative label distribution in the divergence estimate, we propose to estimate a re-weighted source domain label distribution similar to the unknown target label distribution. Furthermore, RAAN estimates the label distribution by minimizing the EM distance during the end-to-end adversarial training procedure. Surprisingly, estimating the label distribution works as an auxiliary task which helps adapt the classifier and also match the marginal feature distribution.

Finally, we evaluated RAAN using a series of experiments considering different domain distribution divergence. This chapter is partially based on earlier work from [8].

5.2 Related Work

In this section, we review the state-of-the-art methods in reducing the domain distribution divergence for the UDA problem.

5.2.1 Matching Feature Distribution Using Adversarial Training

As an unbiased divergence measurement, MMD [19] has been utilized in various Deep Convolutional Neural Network (DCNN) based adaptation networks for matching joint label and feature distribution [30–32, 42, 44, 49]. Most recently, Deep Adaptive Hashing (DAH) applied Multi-Kernel Maximum-Mean-Discrepancy (MK-MMD) for UDA, where the features are projected to the Hamming space for distribution matching [46]. Jenson Shannon (JS)-divergence based adversarial training based methods are currently the best-performing techniques [6, 26, 27, 37, 45]. Ganin et al. [16] first considered the use of a gradient reversal layer for domain-invariant adversarial training, which is further extended by [17, 45, 52] by adding extra loss functions to regularize the target domain features. Recent approaches investigate the OT formulation to match feature or joint label and feature distribution [4, 8, 10, 11, 38]. Similarly, Haeusser et al. [22] proposed the associative method to learn the cross-domain transition map. French et al. [14] proposed the self-ensemble technique, inspired by the semi-supervised method. However, these methods are limited by the required large batch size, sampling from balanced classes and the complex image augmentation techniques.

Inspired by the good performance of generative adversarial networks (GAN) [18], a number of works were proposed to translate image styles between domains directly [6, 39, 43]. Most recent GAN based methods also proposed to transfer style from source to target domain and back again [27, 37], however these pixel-level image translation methods require more investigation when adapting domains having large discrepancy, for example, from the SVHN to the MNIST dataset.

5.2.2 Matching Feature Distribution Using OT

The most closely related approach to RAAN are [10, 11], which both solve the OT directly for reducing the cross-domain distribution divergence. However, these implementations have not been embedded into the end-to-end learning framework and used the stand-alone De-Caffe features [13]. Instead, RAAN utilizes the domain discriminator network with the objective of minimizing the dual formulation of the EM distance. From this point of view, the Wasserstein GAN [2, 20] is a special case to minimize the dual of EM distance with the aim to generate the images from another dataset. To the best of author's knowledge, RAAN may be the first to learn domain invariant features for UDA utilizing the OT based EM distance in deep networks and within the end-to-end pipeline. Note that the concurrent work [38] also adopted the EM distance as the divergence measurement in UDA, however, we are handling the more generalized scenario with unbalanced datasets using the re-weighting scheme and label distribution matching.

5.2.3   Instance Re-weighting Scheme

The instance re-weighting scheme or label distribution matching methods are nothing new in theoretical UDA [9, 49, 50] and or in the causal inference regime [51]. For UDA tasks using deep adaptation methods, only Yan et al. [49] recently proposed to learn the bias of the source domain instances by the classification expectation maximization (CEM) algorithms using the MMD as the divergence measure [7]. In contrast, RAAN differs from [49] since RAAN proposes to estimates the cross-domain density ratio vector of label distributions by embedding its optimization into the adversarial training via back-propagation. Finally, in this chapter, we also argue and explain why matching the label distribution helps to adapt the classifier.

5.3   Method

Suppose we are given two n_{cls}-class domain sets, composed of the source domain $D_s = \{(x_i^s, y_i^s)\}_{i=1}^{n_s}$ and the unlabelled target domain $D_t = \{(x_j^t)\}_{j=1}^{n_t}$, where n_s and n_t are image numbers and y_i^s indicates the source domain label of the image x_i^s.

The network architecture of RAAN is illustrated in Fig. 5.1, including four networks, specifically two conventional L-layer deep network encoders $\mathcal{T}_s$ and $\mathcal{T}_t$, the classifier $\mathcal{C}$ and a domain discriminator network $\mathcal{D}$. The random variables representing the image and label in general are denoted as X and Y.

The first objective of RAAN is to adapt the classifier to target domain, which is difficult without the target domain labels. However, as the label is a low-dimensional and discrete variable, matching their distributions is easier than the one of the high-dimensional features. We also argue that this procedure can assist with the adaptation of classifiers (see the reasons in Sect. 3.2). Based on this assumption and intuition, a re-weighted source domain label distribution $P^{Re}(Y^s)$ is estimated by mapping a variable $\alpha \in \mathbb{R}^{n_{cls}}$ by the soft-max function and the estimation is based on matching the re-weighted source label distribution $P^{Re}(Y^s)$ with that of the unknown target $P^t(Y^t)$. Consequently, the density ratio vector can be denoted as $\beta \in \mathbb{R}^{n_{cls}}$, with its $(y^s)^{th}$ element $\beta(y^s)$ calculated by $\beta(y^s) = \frac{P^{Re}(Y^s=y^s)}{P^s(Y^s=y^s)}$. As β

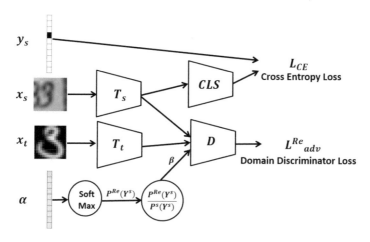

Fig. 5.1 RAAN's architecture: first, in the source domain, the DCNN $\mathcal{T}_s$ and the classifier CLS are trained to extract discriminative features from images x_s labeled by y_s by minimizing the cross entropy loss $\mathcal{L}_{CE}$. Second, to adapt the classifier by matching the label distribution between domains, the re-weighted source domain label distribution $P^{Re}(Y^s)$ is computed by transforming a variable α using the soft-max function. Then it is straightforward to obtain the ratio vector as follows: $\beta = \frac{P^{Re}(Y^s)}{P^s(Y^s)}$. To extract transferable features for target domain images x^t, the target domain DCNN $\mathcal{T}_t$, domain discriminator $\mathcal{D}$ and the estimated density ratio vector β play the following adversarial game: β and $\mathcal{D}$ tries to discriminate whether features is from the target or source domain, while $\mathcal{T}_t$ tries to confuse $\mathcal{D}$ and β

can be calculated directly based on α, in the following chapter, we regard β as the variable under estimate.

The second objective of RAAN is to learn the domain-invariant and discriminative marginal features $P^t(\mathcal{T}_t^{l=L}(X^t))$ and $P^s(\mathcal{T}_s^{l=L}(X^s))$ by reducing their distribution divergence. For brevity and since this chapter only considers matching the last convolution layer's features, we denote $\mathcal{T}_t$ and $\mathcal{T}_s$ to replace $\mathcal{T}_t^{l=L}$ and $\mathcal{T}_s^{l=L}$ respectively in the following chapter. Given the source domain images and labels $\{x^s, y^s\} \in D_s$, the encoder $\mathcal{T}_s$ is utilized to extract discriminative features $\mathcal{T}_s(x^s)$ for the classification task, where we jointly train the classifier CLS and $\mathcal{T}_s$ by minimizing the cross-entropy loss $\mathcal{L}_{CE}$ as follows:

$$\min_{\mathcal{T}_s, CLS} \mathcal{L}_{CE}. \tag{5.1}$$

To obtain transferable features $\mathcal{T}_t(x^t)$ without labels, $\mathcal{T}_t$ plays an adversarial game with the domain discriminator network $\mathcal{D}$ and the ratio vector β so that the divergence between the re-weighted feature distribution in the source domain $\beta(y^s)P^s(\mathcal{T}_s(x^s))$ and the target domain $P^t(\mathcal{T}_t(x^t))$ is reduced. Additionally, to reduce the potentially disparate cross-domain divergence, the OT-based EM distance is reformulated in the adversarial manner, with more details shown in Sect. 3.1.

Specifically, RAAN is trained in the following adversarial manner, where the $\mathcal{T}_t$ wants to confuse discriminator network $\mathcal{D}$ no matter how hard $\mathcal{D}$ and β try to differentiate the features from two domains. Based on the discriminator loss $\mathcal{L}_{adv}^{Re}$, the following objective function can be obtained:

$$\min_{\mathcal{T}_t} \max_{\mathcal{D},\beta} \mathcal{L}_{adv}^{Re}. \tag{5.2}$$

In fact, besides helping the adaptation of the classifier, matching the label distributions also eases the difficulty of matching the marginal feature distribution. The possible reason may be: if we assume that the feature generation processes are the same between domains, that is $P^s(\mathcal{T}^s(X^s)|Y^s) = P^t(\mathcal{T}^t(X^t)|Y^t)$, then $P^{Re}(Y^s) = P^t(Y^t)$ helps match the marginal feature distributions $P^s(\mathcal{T}^s(X^s)) = P^t(\mathcal{T}^t(X^t))$.

Section 3.1 introduces the OT based EM distance and its implementation to match the marginal cross-domain feature distributions in an adversarial manner. Then Sect. 3.2 focuses on the proposed label distribution matching module and the way to embed it in the adversarial training. We also provide insights of why this label distribution matching module helps to adapt the classifier and meanwhile to match marginal feature distributions. Finally in Sect. 3.3, we summarize the final objective function of RAAN.

5.3.1 Optimal Transport in Adversarial Training

Formally, the empirical distributions of source and target marginal features μ^s and μ^t can be formulated in the following respectively:

$$\mu^s = \sum_i^{n_s} p_i^s \delta_{\mathcal{T}_s(x_i^s)}, \mu^t = \sum_j^{n_t} p_j^t \delta_{\mathcal{T}_t(x_j^t)} \tag{5.3}$$

where $\delta_{\mathcal{T}_s(x_i^s)}$ and $\delta_{\mathcal{T}_t(x_j^t)}$ are the Dirac functions at location $\mathcal{T}_s(x_i^s)$ and $\mathcal{T}_t(x_j^t)$ and p_i^s and p_j^t are their probability masses.

Ideally, the cross-domain distribution matching problem can be modelled by the joint probabilistic coupling, or the transportation plan between feature distributions in source and target domains and OT is proposed to calculate such an optimal transportation plan. If we denote the coupling under estimate as γ with the marginals μ^s and μ^t, the set of probabilistic couplings B can be defined as the following [11]:

$$B = \left\{ \gamma \in (\mathbb{R}^+)^{n_s \times n_t} | \gamma \mathbf{1}_{n_t} = \mu^s, \gamma^T \mathbf{1}_{n_s} = \mu^t \right\}. \tag{5.4}$$

In general, to reduce feature distribution divergence, OT based adaptation methods first measure the optimal transportation plan between two distributions and then learn the features to minimize the cost of such a plan. Therefore, we first define the metric $J(\mu^s, \mu^t)$ in Eq. (5.5) to measure the total cost of transporting probability masses from target to source domains,

$$J(\mu^s, \mu^t) = \langle \gamma, C \rangle_F, \quad with \; \gamma \in B, \tag{5.5}$$

where C is the distance matrix whose $(i, j)^{th}$ element is defined by the distance cost function $c(\mathcal{T}_s(x_i^s), \mathcal{T}_t(x_j^t))$ between features. The $(i, j)^{th}$ element $\gamma(i, j)$ indicates how much mass is moved from $\mathcal{T}_t(x_j^t)$ to $\mathcal{T}_s(x_i^s)$, and F is the Frobenius dot product. Subsequently for brevity, we drop the index i, j to represent x_i^s, x_j^t as x^s, x^t.

After that, the OT γ_0 can be estimated by minimizing the cost $J(\mu^s, \mu^t)$ in Eq. (5.6), with the optimal transportation cost or the well-known EM distance defined by $W(\mu^s, \mu^t)$ in Eq. (5.7) [24]. Finally, assuming the ideal source domain features $\mathcal{T}_s(x^s)$ are available, to learn the transferable features in target domains, it is intuitive to train the DCNN transformation $\mathcal{T}_t$ under the objective of minimizing the EM distance $W(\mu^s, \mu^t)$, as shown in Eq. (5.8) [11].

$$\gamma_0 = \underset{\gamma \in B}{\arg\min} \, J(\mu^s, \mu^t) \tag{5.6}$$

$$W(\mu^s, \mu^t) = \underset{\gamma \in B}{\min} \, J(\mu^s, \mu^t) \tag{5.7}$$

$$\underset{\mathcal{T}_t}{\min} \, W(\mu^s, \mu^t) \tag{5.8}$$

Conventional methods to compute the constraint of γ in Eq. (5.4) requires either linear programs or iterative algorithms which limits the end-to-end learning in some sense. Therefore, we utilized the dual formulation of $W(\mu^s, \mu^t)$ in Eqs. (5.9) and (5.10) (following equation (5.3) in [47]), considering the capability of batch-wise back-propagation. More specifically, we use the domain discriminator network $\mathcal{D}$ and its variant $\hat{\mathcal{D}}$ as two dual functions in the following:

$$W(\mu^s, \mu^t) = \underset{\mathcal{D}, \; \hat{\mathcal{D}}}{\max} \, \mathcal{L}_{adv}, \quad where$$

$$\mathcal{L}_{adv} = \underset{x^s \sim P^s(X^s)}{\mathbb{E}} \mathcal{D}(\mathcal{T}_s(x^s)) + \underset{x^t \sim P^t(X^t)}{\mathbb{E}} \hat{\mathcal{D}}(\mathcal{T}_t(x^t)) \tag{5.9}$$

$$s.t. \mathcal{D}(\mathcal{T}_s(x^s)) + \hat{\mathcal{D}}(\mathcal{T}_t(x^t)) \leq c(\mathcal{T}_s(x^s), \mathcal{T}_t(x^t)). \tag{5.10}$$

In this paper, we choose the following distance cost function $c(\mathcal{T}_s(x^s), \mathcal{T}_t(x^t)) = \|\mathcal{T}_s(x^s) - \mathcal{T}_t(x^t)\|$ for reasons of computational efficiency and permitting gradient measurements, however, this does not infer that it is the only possible function that can be selected. According to the constraint (10), the best function that $\hat{\mathcal{D}}$ has to be is $-\mathcal{D}$, as $c(\mathcal{T}_s(x^s), \mathcal{T}_t(x^t))$ is defined to be non-negative. In this way, the constraint in (5.10) is equivalent to ensuring that $\mathcal{D}$ is a 1-Lipschitz function, or alternatively its gradient norm is smaller than 1. Therefore, if we use (5.9) and (5.10) in (5.8) to replace the EM distance, the DCNN transformation $\mathcal{T}_t$ and the domain discriminator network $\mathcal{D}$ can be trained based on the mini-max objective function in (5.11),

$$\min_{\mathcal{T}_t} W(\boldsymbol{\mu}^s, \boldsymbol{\mu}^t) = \min_{\mathcal{T}_t} \max_{\mathcal{D}} \mathcal{L}_{adv}, \; where$$

$$\mathcal{L}_{adv} = \sum_{(x^s, y^s) \sim P^s(X, Y^s)} \mathcal{D}(\mathcal{T}_s(x^s)) P^s(\mathcal{T}_s(x^s)|y^s) P^s(y^s)$$

$$- \mathop{\mathbb{E}}_{x^t \sim P^t(X^t)} \mathcal{D}(\mathcal{T}_t(x^t))$$

$$s.t. \| \nabla_{\mathcal{T}_t(x^t)} \mathcal{D}(\mathcal{T}_t(x^t)) \|_2 \leq 1,$$

$$\| \nabla_{\mathcal{T}_s(x^s)} \mathcal{D}(\mathcal{T}_s(x^s)) \|_2 \leq 1. \tag{5.11}$$

5.3.2 Adapting the Classifier by Label Distribution Matching

Although OT based EM distance is utilized to match marginal feature distributions $P^s(\mathcal{T}_s(X^s))$ and $P^t(\mathcal{T}_t(X^t))$, it cannot help the classifier adaptation, since $P^s(\mathcal{T}_s(X^s)) = P^t(\mathcal{T}_t(X^t))$ does not infer $P^s(Y^s|\mathcal{T}_s(X^s)) = P^t(Y^t|\mathcal{T}_t(X^t))$. However, according to Bayes rule in (5.12), instead of matching $P^t(Y^t|\mathcal{T}_t(X^t))$ and $P^s(Y^s|\mathcal{T}_s(X^s))$ directly, we can learn $\mathcal{T}_t$ under the objective of matching $P^s(\mathcal{T}_s(X^s)|Y^s) P^s(Y^s)$ and $P^t(\mathcal{T}_t(X^t)|Y^t) P^t(Y^t)$,

$$P(\mathcal{T}(X)|Y)P(Y) \propto P(Y|\mathcal{T}(X)). \tag{5.12}$$

Without any prior knowledge of the target label distribution $P^t(Y^t)$, it is nontrivial to directly match $P^t(\mathcal{T}_t(X^t)|Y^t)P^t(Y^t)$ and $P^s(\mathcal{T}_s(X^s)|Y^s)P^s(Y^s)$. However, inspired by the fact that the label is a low-dimensional and discrete variable with well-defined distribution, it is more straightforward to match the cross-domain label distributions compared with its conditional variant. This inspires us further to investigate whether matching label distributions, part of $P^t(\mathcal{T}_t(X^t)|Y^t)P^t(Y^t)$ and $P^s(\mathcal{T}_s(X^s)|Y^s)P^{Re}(Y^s)$ helps the classifier adaptation. More specifically, we propose to estimate the re-weighted source domain label distribution $P^{Re}(Y^s)$ so that it can match the unknown $P^t(Y^t)$ in the target domain.

Since the EM-distance has been adopted to match $P^t(\mathcal{T}_t(X^t))$ and $P^s(\mathcal{T}_s(X^s))$, we integrate the re-weighted label distribution $P^{Re}(Y^s)$ into the procedure of matching the *re-weighted* marginal feature distributions $P^s(\mathcal{T}_s(X^s))$ and $P^t(\mathcal{T}_t(X^t))$ in the adversarial training. To guarantee the validity of the learned re-weighted label distribution $P^{Re}(Y^s)$, we constrain the following conditions utilizing the softmax function in the implementation:

$$\sum_{i=1}^{n_{cls}} P^{Re}(Y^s = y_i) = 1, \tag{5.13}$$

where y_i indicates the label of the ith class.

Finally, we directly choose to replace the $P^s(Y^s)$ by $P^{Re}(Y^s)$ in the mini-max objective function $\mathcal{L}_{adv}$ in (11) and optimize the network based on the new adversarial training objective $\mathcal{L}_{adv}^{Re}$ in (14), where the network $\mathcal{D}$, $\mathcal{T}_t$ and the ratio vector $\boldsymbol{\beta}$ are trained in the following manner: $\mathcal{D}$ and $\boldsymbol{\beta}$ are trained in a cooperative way to estimate the EM-distance, while $\mathcal{T}_t$ is trained to minimize the EM-distance. From another perspective to understand the rationale behind $\mathcal{L}_{adv}^{Re}$, $\boldsymbol{\beta}$ is utilized to assign the importance weights to different classes' images $\boldsymbol{x}^s$ in the source domain, so that the mini-batches in the two domains are sampled from the similar distributions. This procedure, to some extent eases the optimization complexity of network $\mathcal{D}$ and $\mathcal{T}_t$ since they can focus on matching $P^s(\mathcal{T}_s(\boldsymbol{x}^s))$ and $P^t(\mathcal{T}_t(\boldsymbol{x}^t))$ without being confused by the unbalanced domain distributions.

$$\min_{\mathcal{T}_t} \max_{\mathcal{D},\boldsymbol{\beta}} \mathcal{L}_{adv}^{Re}, \; where$$

$$\mathcal{L}_{adv}^{Re} = \sum_{(\boldsymbol{x}^s,y^s)\sim P^s(X^s,Y^s)} \mathcal{D}(\mathcal{T}_s(\boldsymbol{x}^s)) P^s(\mathcal{T}_s(\boldsymbol{x}^s)|y^s) P^{Re}(y^s)$$

$$- \mathop{\mathbb{E}}_{\boldsymbol{x}^t\sim P^t(X^t)} \mathcal{D}(\mathcal{T}_t(\boldsymbol{x}^t))$$

$$= \sum_{(\boldsymbol{x}^s,y^s)\sim P^s(X^s,Y^s)} \mathcal{D}(\mathcal{T}_s(\boldsymbol{x}^s)) P^s(\mathcal{T}_s(\boldsymbol{x}^s)|y^s) \boldsymbol{\beta}(y^s) P^s(y^s)$$

$$- \mathop{\mathbb{E}}_{\boldsymbol{x}^t\sim P^t(X^t)} \mathcal{D}(\mathcal{T}_t(\boldsymbol{x}^t))$$

$$= \mathop{\mathbb{E}}_{(\boldsymbol{x}^s,y^s)\sim P^s(X^s,Y^s)} \boldsymbol{\beta}(y^s)\mathcal{D}(\mathcal{T}_s(\boldsymbol{x}^s)) - \mathop{\mathbb{E}}_{\boldsymbol{x}^t\sim P^t(X^t)} \mathcal{D}(\mathcal{T}_t(\boldsymbol{x}^t))$$

$$s.t. \| \nabla_{\mathcal{T}_t(\boldsymbol{x}^t)} \mathcal{D}(\mathcal{T}_t(\boldsymbol{x}^t))\|_2 \leq 1,$$

$$\| \nabla_{\mathcal{T}_s(\boldsymbol{x}^s)} \mathcal{D}(\mathcal{T}_s(\boldsymbol{x}^s))\|_2 \leq 1. \tag{5.14}$$

5.3.3 Optimization in RAAN

As shown in Fig. 5.1, RAAN is proposed to jointly minimize the cross entropy loss of the source domain samples and to reduce the cross-domain divergence of the extracted feature distributions. The following equation illustrates the empirical estimate of the re-weighted adversarial loss $\mathcal{L}_{adv}^{Re}$:

$$\mathcal{L}_{adv}^{Re} = \frac{1}{n_s} \sum_{i=1}^{n_s} \mathcal{D}(\boldsymbol{\beta}(y_i^s) \mathcal{T}_s(\boldsymbol{x}_i^s)) - \frac{1}{n_t} \sum_{j=1}^{n_t} \mathcal{D}(\mathcal{T}_t(\boldsymbol{x}_j^t)). \tag{5.15}$$

Following on from the idea of controlling the 1-Lipschitz function of the domain discriminator network $\mathcal{D}$ [20], we explicitly constrain the gradient norm penalty term as follows:

$$\mathcal{L}_{gp} = \| \nabla_{\hat{\mathcal{T}}(\hat{\boldsymbol{x}})} \mathcal{L}_{adv}^{Re} - 1 \|_2, \tag{5.16}$$

where $\hat{\mathcal{T}}(\hat{\boldsymbol{x}})$ is the weighted interpolation samples of $\mathcal{T}_t(\boldsymbol{x}^t)$ and $\mathcal{T}_s(\boldsymbol{x}^s)$. In summary, the total objective function in RAAN is formulated in the following adversarial manner:

$$\min_{\mathcal{T}_t, \mathcal{D}, \boldsymbol{\beta}} -\mathcal{L}_{adv}^{Re} + \lambda_{gp} \mathcal{L}_{gp} + \lambda_{reg} \|\boldsymbol{\beta}\|_2, \tag{5.17}$$

$$\min_{\mathcal{T}_t} -\frac{1}{n_t} \sum_{j=1}^{n_t} \mathcal{D}(\mathcal{T}_t(\boldsymbol{x}_j^t)), \tag{5.18}$$

$$\min_{\mathcal{T}_s, CLS} \frac{1}{n_s} \sum_{i=1}^{n_s} \mathcal{L}_{CE}(CLS(\mathcal{T}_s(\boldsymbol{x}_i^s)), y_i^s), \tag{5.19}$$

where $\mathcal{L}_{CE}(CLS(\mathcal{T}_s(\boldsymbol{x}_i^s)), y_i^s)$ indicates the cross-entropy function with classifier CLS, feature vector $\mathcal{T}_s(\boldsymbol{x}_i^s)$ and its label y_i^s. Note that to train the networks stably, the source domain DCNN $\mathcal{T}_s$ is trained first while $\mathcal{T}_t$ and $\mathcal{D}$ are trained to match the feature distributions between $\mathcal{T}_t(\boldsymbol{x}^t)$ and $\mathcal{T}_s(\boldsymbol{x}^s)$ in an adversarial manner. In addition, to stably learn the ratio vector, we add the L2-norm of $\boldsymbol{\beta}$ as the regularization term in (5.17). λ_{gp} and λ_{reg} indicate the regularization weights of the gradient penalty term and the L2-norm of the ratio vector respectively.

5.4 Experiment and Results

This section evaluates RAAN using two UDA tasks, specifically one between handwritten digit datasets and the other between cross-modality datasets. For all the experiments, RAAN achieves competitive results compared with the state-of-the-art methods and outperforms them by a large margin when the distribution divergence is large between domains.

5.4.1 Adaptation Tasks and Dataset

The first UDA task adapts among four hand written digit datasets including MNIST [25], USPS [12], SVHN [34] and MNIST-M [15]. As shown in Fig. 5.2, adaptation between these four datasets are of varying difficulty. MNIST and USPS are grey-scale image datasets collected in a fairly well-controlled environment while images in MNIST-M are synthesized using the patches from the BSDS500 dataset [1] as the background and the MNIST images as the foreground. To evaluate RAAN in reducing large domain discrepancies, we use SVHN in the adaptation tasks, which is composed of RGB images in more complicated real-world scenarios, e.g. misalignment of images and different light conditions. In addition, note that the sub-class instances between SVHN and the others are largely unbalanced.

To continue evaluating RAAN in reducing large domain shifts, the second adaptation task is designed using the NYU-D dataset [40], adapting from the indoor object images in RGB format to the depth variants encoded by the HHA format [21]. The 19-class dataset is extracted following the scheme in [45]. As shown in Fig. 5.2, the domain shifts between images of RGB and HHA format are fairly large, mainly due to the low image resolutions and potential mis-alignments caused by the coarse cropping box. In addition, as shown in the instance number in Table 5.4, this dataset has unbalanced sub-class instances. Furthermore, it is challenging as the images from the target domain are collected through another sensor in a completely different format from the source images.

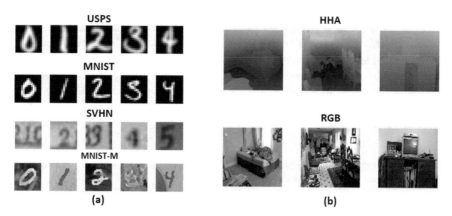

Fig. 5.2 DA datasets (**a**) four hand-written digit datasets; (**b**) cross-modality dataset including RGB and RGB-depth images

5.4.2 Adaptation in Hand-Written Digit Dataset

For the task of adapting between hand-written digit datasets, the following four adaptation directions are chosen for the evaluation: MNIST $\rightarrow$ USPS, USPS $\rightarrow$ MNIST, SVHN $\rightarrow$ MNIST and MNIST $\rightarrow$ MNIST-M. For the first three tasks, the base encoder network is a variant of LeNet denoted as $\mathcal{T}_s$, $\mathcal{T}_t$ and the domain discriminator network $\mathcal{D}$ is composed of three fully-connected layers activated by the rectified linear unit (ReLU) with output activation numbers of 512, 512, 1 respectively. For adapting from MNIST to MNIST-M, we adopt the basic model architecture of pixelDA [6] but change the objective to an re-weighted OT-based objective function. We followed the experiment protocol in [28, 45] for the adaption tasks. The protocol used for MNIST $\rightarrow$ MNIST-M is the same as that in [6] for fair comparisons.

To clearly evaluate the performance of RAAN, we denote RAAN(+) and RAAN(−) as RAAN with and without the re-weighting scheme respectively. As shown in Table 5.1, when adapting between MNIST and USPS, compared with ADDA and Co-GAN, the proposed RAAN(−) and RAAN(+) achieved competitive results and RAAN(+) slightly outperforms RAAN(−). However, in the task SVHN $\rightarrow$ MNIST, RAAN(−) and RAAN(+) achieved 80.7% and 89.1% respectively, outperforming the state-of-the-art ADDA by 4.7% and 13.1% respectively, while Co-GAN does not converge in this experiment. It may suggest that facing two disparate domains, the weight-sharing approach used in Co-GAN is not capable of bridging the cross-domain discrepancies by simply transferring styles of domain datasets such as MNIST and SVHN. As RAAN(−) utilized the same DCNN architecture to ADDA, RAAN(−)'s superior performance is mainly due to the OT based objective function. We hypothesize that the OT based EM distance is able to better reduce feature distribution divergence of two disparate domains, e.g., SVHN and MNIST. In addition, based on the fact that RAAN(+) outperforms both ADDA and RAAN(−), we hypothesize that matching the label distribution helps adapt the classifiers, which also helps to minimize the EM distance of marginal feature distributions.

Table 5.1 Recognition rates of adapting hand-written digit datasets; RAAN(+) and RAAN(−) indicate with and without the re-weighting scheme

Methods	MNIST to USPS	USPS to MNIST	SVHN to MNIST
Source only	0.725	0.612	0.593
Gradient reversal[16]	0.771	0.730	0.739
Domain confusion [44]	0.791	0.665	0.681
Co-GAN[26]	**0.912**	0.891	No converge
ADDA [45]	0.894	0.901	0.760
RAAN(−) (Ours)	0.883	0.915	0.807
RAAN(+) (Ours)	0.89	**0.921**	**0.892**

Table 5.2 Recognition rates of adapting from MNIST to MNIST-M; RAAN(+) and RAAN(−) indicate with and without re-weighting scheme

Dataset	Source only [6]	CORAL [42]	MMD [6]	DANN [16]	DSN [5]	PixelDA [6]	RAAN(+)/(−) (Ours)
MNIST to MNIST-M	0.636	0.577	0.769	0.774	0.832	0.982	**0.985**

Table 5.3 $\mathcal{A}$-Distance of adversarial training method

Metric	Source only	ADDA	RAAN(−)	RAAN(+)
$\mathcal{A}$-Distance	1.673	1.548	1.526	1.506

For the task MNIST → MNIST-M, as shown in Table 5.2, RAAN achieves slightly better performance than pixelDA. In addition, as expected RAAN(−) and RAAN(+) achieve similar results since these two dataset don't have severely unbalanced label distributions. It has been argued and widely recognized that domain shift between MNIST and MNIST-M is large for a conventional deep network based method [16]. However, in this section, we argue that reducing the domain shift caused by the background images in MNIST-M is easier than reducing the one between MNIST and SVHN .

5.4.3 Adaptation in Cross-modality Dataset

In this section, we evaluated RAAN in the presence of large domain shifts that confront the adaptation from RGB images to RGB-depth images. To enable a fair comparison, we follow ADDA's experimental set-up [45] and utilized the VGG-16 architecture [41] for DCNNs $\mathcal{T}_s$ and $\mathcal{T}_t$. The domain discriminator network $\mathcal{D}$ is composed of three fully-connected layers activated by the Relu, with 1024, 2048, 1 outputs respectively (Table 5.3).

As shown in Table 5.4, the sub-class classification accuracies are reported using RAAN(−) and RAAN(+), along with the re-weighted label distribution $P^{Re}(Y^s)$ yielded by RAAN(+) and the target one $P^t(Y^t)$. It can be observed from the overall recognition rates that RAAN(+) achieves an average of 34.3%, outperforming ADDA by 6.7% and RAAN(−) by 3.5%. In addition, RAAN(−) outperforms ADDA by 3.2%. For classes with fewer samples, RAAN(+) and RAAN(−) achieve better performances and ADDA only achieved better performance in class 'chair' as that class has the largest number of samples. It can also be observed that RAAN(+) outperforms RAAN(−) not only from the overall recognition accuracy but also in terms of how many classes the classifier can recognize (i.e., classes with the recognition rates more than 0%). It can also be observed that the re-weighting scheme increases the importance weights of instances from the sub-classes with a lower number of instances. This can be verified by comparing the number of sub-class instances with the estimated ratio vector β in Table 5.4.

Table 5.4 Adaptation results (%) in cross-modality dataset; RAAN(+) and RAAN(-) indicate with and without re-weighting scheme

class	bathtub	bed	bookshelf	box	chair	counter	desk	door	dresser	garbage bin	lamp	monitor	night stand	pillow	sink	sofa	table	television	toilet	Overall
NO.instances	19	96	87	210	611	103	122	129	25	55	144	37	51	276	47	129	210	33	17	2401
Source only	0.0	76.0	0.0	10.5	13.1	2.9	9.0	45.7	0.0	3.6	12.5	0.0	0.0	54.9	0.0	2.3	14.8	3.0	0.0	18.9
ADDA	0.0	46.9	0.0	0.5	**76.2**	19.4	1.6	51.9	4.0	1.8	0.7	0.0	0.0	8.3	0.0	6.2	13.8	0.0	0.0	27.6
RAAN(-) (ours)	**0.000**	**0.500**	**0.299**	0.0	62.9	1.9	0.0	27.1	0.0	0.0	0.0	0.0	0.0	**69.2**	0.0	0.0	4.3	0.0	0.0	30.8
RAAN(+) (ours)	0.0	10.4	0.0	**14.8**	70.3	**48.5**	0.0	**61.2**	0.0	0.0	0.7	0.0	0.0	63.8	**10.6**	0.0	**20.5**	0.0	0.0	**34.3**
$Est.\beta$	2.040	1.395	1.568	0.872	0.492	1.440	1.297	1.206	1.640	1.629	1.057	1.858	1.652	0.805	1.527	1.102	0.682	1.814	1.858	---

5.4.4 Parameter Selection and Implementation

The experiments are conducted on a GPU cluster. For all experiments, we utilize the Adam optimizer, with the learning rate selected using the parameter sweep in the range of $(2e^{-5}, 1e^{-2})$. For the regularization weights, λ_{gp} and λ_{reg} are chosen based on parameter sweep results in the ranges of $(1, 100)$ and $(0.01, 100)$ respectively. We used the exponential decay, with decay factor of 0.99 for every 1000 iterations. All experiments are run 10 times, each for 100,000 iterations and we report the average results. For adapting from MNIST to MNIST-M, the batch size is 32 while for others, the batch size is 128.

5.5 Analysis

In this section, we analyze the re-weighting scheme to match the label distribution and the proposed OT based distribution matching machine in both quantitative and qualitative ways. The evaluation is in the context of the most challenging scenario involving the adaptation from the SVHN to the MNIST.

5.5.1 Evaluate the Re-weighting Scheme

Figure 5.3 illustrates the comparison between ground truth label distribution ratio vector (red) and the learned one (blue). It can be seen that some ratios are accurate while others are not. However, the relative ratio trend of the learned ratio vector β follows that of the ground truth.

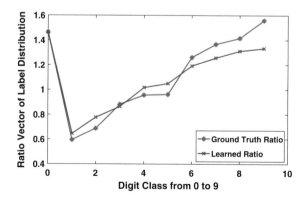

Fig. 5.3 Ratio of label distribution between SVHN and MNIST; red line indicates the ground truth ratio, while blue one indicates the estimated ratio

Class imbalance problem widely exists in the real-world application datasets and such mismatched label distributions between SVHN and MNIST confuse the domain discriminator and the encoder learning where the feature distributions will be matched in a biased manner. Through the discussions in Sect. 3.2, the mismatch of label distribution will directly result in the mismatch of classifiers as well. However, as shown in Fig. 5.3, RAAN(+) is able to successfully match the label distributions by simply estimating the ratio vector embedded in the adversarial training. Therefore, this can be regarded as the main reason for the 9% improvement achieved by RAAN(+) compared to RAAN(−) shown in Table 5.1. To sum up, matching the label distribution can adapt the classifiers.

We can understand the re-weighted distribution matching in the following way: the module works to assign different importance weights to source domain instances but with the constraints that the instances from the same class have the same weight. For example, as shown in Fig. 5.3, the learned ratio of digit "0" is around 1.5, which means that in the adversarial training, each sample from digit "0" in SVHN dataset can be regarded as 1.5 samples.

5.5.2 Evaluate Distribution Divergence of Feature Embeddings

To analyze the distribution divergence in a quantitative way, we calculate the $\mathcal{A}$ distance suggested by the UDA community [3], taking the input features extracted by adaptation models under comparison. We utilized the SVM classifier to calculate the generalization error θ of classifying the source and target domain features as a binary classification task. Then the $\mathcal{A}$ distance is as follows: $d = 2(1 − 2\theta)$. As shown in Table 5.3, the $\mathcal{A}$ distances of feature embeddings with no adaptation, adapted by ADDA, OT based RAAN(−) and RAAN(+) progressively decrease. Since RAAN uses the same deep network architecture as ADDA's, compared with ADDA, RAAN(−) achieved the lower $\mathcal{A}$ distance inferring the fact that feature distribution between domains can be better matched using RAAN(−). This further proves the fact that the OT based EM distance is a better measure to estimate and reduce the large distribution divergence than the geometry-oblivious JS divergence. In addition, compared to RAAN(−), the smaller $\mathcal{A}$ distance achieved by RAAN(+) further indicates that matching the label distribution is the cooperative task of matching the marginal feature distribution.

Finally, to measure the feature distribution divergence in a qualitative way, we utilized the T-SNE software package [33] to visualize the 2-D embedding of the extracted features. In Fig. 5.4, the example feature points from the same class adapted by RAAN in Fig. 5.4c are clustered closer than those in Fig. 5.4b by ADDA and also those without the adaptation method in Fig. 5.4a.

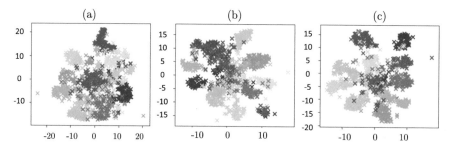

Fig. 5.4 T-SNE plot of features when adapting from SVHN to MNIST; (**a**) no adaptation; (**b**) adaptation after ADDA; (**c**) adaptation after RAAN. We randomly select 1000 features samples from 10 classes, with 100 samples per class

5.6   Conclusions

In this chapter, we propose a Re-weighted Adversarial Adaptation Network (RAAN) to reduce disparate domain feature distribution and adapt the classifier. Through extensive experiments, RAAN has been shown to outperform others when the cross-domain distribution divergence is large and disparate. The superior performance of adapting the domains with unbalanced class achieved by RAAN also reveals the fact that integrating the label ratio vector estimation and label distribution matching in adversarial training can not only adapt the classifier, but also help match the marginal feature distribution. Future work will consider comparing the primal formulation of OT and reformulate this in the end-to-end training.

References

1. Arbelaez, P., Maire, M., Fowlkes, C., Malik, J.: Contour detection and hierarchical image segmentation. IEEE Trans. Pattern Anal. Mach. Intell. **33**(5), 898–916 (2011)
2. Arjovsky, M., Chintala, S., Bottou, L.: Wasserstein generative adversarial networks. In: International Conference on Machine Learning, pp. 214–223 (2017)
3. Ben-David, S., Blitzer, J., Crammer, K., Kulesza, A., Pereira, F., Vaughan, J.W.: A theory of learning from different domains. Mach. Learn. **79**(1), 151–175 (2010)
4. Bhushan Damodaran, B., Kellenberger, B., Flamary, R., Tuia, D., Courty, N.: Deepjdot: deep joint distribution optimal transport for unsupervised domain adaptation. In: The European Conference on Computer Vision (ECCV) (2018)
5. Bousmalis, K., Trigeorgis, G., Silberman, N., Krishnan, D., Erhan, D.: Domain separation networks. In: Advances in Neural Information Processing Systems, pp. 343–351 (2016)
6. Bousmalis, K., Silberman, N., Dohan, D., Erhan, D., Krishnan, D.: Unsupervised pixel-level domain adaptation with generative adversarial networks. In: Proceedings of the IEEE Conference on Computer Vision and Pattern Recognition, pp. 3722–3731 (2017)
7. Celeux, G., Govaert, G.: A classification EM algorithm for clustering and two stochastic versions. Computat. Stat. Data Anal. **14**(3), 315–332 (1992)

8. Chen, Q., Liu, Y., Wang, Z., Wassell, I., Chetty, K.: Re-weighted adversarial adaptation network for unsupervised domain adaptation. In: Proceedings of the IEEE Conference on Computer Vision and Pattern Recognition, pp. 7976–7985 (2018)

9. Cortes, C., Mansour, Y., Mohri, M.: Learning bounds for importance weighting. In: Advances in Neural Information Processing Systems, pp. 442–450 (2010)

10. Courty, N., Flamary, R., Habrard, A., Rakotomamonjy, A.: Joint distribution optimal transportation for domain adaptation. In: Advances in Neural Information Processing Systems, pp. 3730–3739 (2017)

11. Courty, N., Flamary, R., Tuia, D., Rakotomamonjy, A.: Optimal transport for domain adaptation. IEEE Trans. Pattern Anal. Mach. Intell. **39**(9), 1853–1865 (2017)

12. Denker, J.S., Gardner, W., Graf, H.P., Henderson, D., Howard, R., Hubbard, W., Jackel, L.D., Baird, H.S., Guyon, I.: Neural network recognizer for hand-written zip code digits. In: Advances in Neural Information Processing Systems, pp. 323–331 (1989)

13. Donahue, J., Jia, Y., Vinyals, O., Hoffman, J., Zhang, N., Tzeng, E., Darrell, T.: DeCaF: a deep convolutional activation feature for generic visual recognition. In: International Conference on Machine Learning, pp. 647–655 (2014)

14. French, G., Mackiewicz, M., Fisher, M.: Self-Ensembling for Visual Domain Adaptation (2018). https://iclr.cc

15. Ganin, Y., Lempitsky, V.: Unsupervised domain adaptation by backpropagation. In: International Conference on Machine Learning, pp. 1180–1189 (2015)

16. Ganin, Y., Ustinova, E., Ajakan, H., Germain, P., Larochelle, H., Laviolette, F., Marchand, M., Lempitsky, V.: Domain-adversarial training of neural networks. J. Mach. Learn. Res. **17**(59), 1–35 (2016)

17. Ghifary, M., Kleijn, W.B., Zhang, M., Balduzzi, D., Li, W.: Deep reconstruction-classification networks for unsupervised domain adaptation. In: European Conference on Computer Vision, pp. 597–613. Springer, New York (2016)

18. Goodfellow, I., Pouget-Abadie, J., Mirza, M., Xu, B., Warde-Farley, D., Ozair, S., Courville, A., Bengio, Y.: Generative adversarial nets. In: Advances in Neural Information Processing Systems, pp. 2672–2680 (2014)

19. Gretton, A., Borgwardt, K.M., Rasch, M.J., Schölkopf, B., Smola, A.: A kernel two-sample test. J. Mach. Learn. Res. **13**, 723–773 (2012)

20. Gulrajani, I., Ahmed, F., Arjovsky, M., Dumoulin, V., Courville, A.C.: Improved training of Wasserstein gans. In: Advances in Neural Information Processing Systems, pp. 5767–5777 (2017)

21. Gupta, S., Girshick, R., Arbeláez, P., Malik, J.: Learning rich features from RGB-D images for object detection and segmentation. In: European Conference on Computer Vision, pp. 345–360. Springer, New York (2014)

22. Haeusser, P., Frerix, T., Mordvintsev, A., Cremers, D.: Associative domain adaptation. In: International Conference on Computer Vision (ICCV), vol. 2, p. 6 (2017)

23. He, K., Zhang, X., Ren, S., Sun, J.: Deep residual learning for image recognition. In: Proceedings of the IEEE Conference on Computer Vision and Pattern Recognition, pp. 770–778 (2016)

24. Kantorovitch, L.: On the translocation of masses. Manage. Sci. **5**(1), 1–4 (1958)

25. LeCun, Y., Bottou, L., Bengio, Y., Haffner, P.: Gradient-based learning applied to document recognition. Proc. IEEE **86**(11), 2278–2324 (1998)

26. Liu, M.Y., Tuzel, O.: Coupled generative adversarial networks. In: Advances in Neural Information Processing Systems, pp. 469–477 (2016)

27. Liu, M.Y., Breuel, T., Kautz, J.: Unsupervised image-to-image translation networks. In: Guyon, I., Luxburg, U.V., Bengio, S., Wallach, H., Fergus, R., Vishwanathan, S., Garnett, R. (eds.) Advances in Neural Information Processing Systems, vol. 30, pp. 700–708. Curran Associates, Inc., Red Hook, NY (2017) http://papers.nips.cc/paper/6672-unsupervised-image-to-image-translation-networks.pdf

28. Long, M., Wang, J., Ding, G., Sun, J., Yu, P.S.: Transfer feature learning with joint distribution adaptation. In: Proceedings of the IEEE International Conference on Computer Vision, pp. 2200–2207 (2013)
29. Long, M., Wang, J., Ding, G., Sun, J., Yu, P.S.: Transfer joint matching for unsupervised domain adaptation. In: Proceedings of the IEEE Conference on Computer Vision and Pattern Recognition, pp. 1410–1417 (2014)
30. Long, M., Cao, Y., Wang, J., Jordan, M.: Learning transferable features with deep adaptation networks. In: International Conference on Machine Learning, pp. 97–105 (2015)
31. Long, M., Zhu, H., Wang, J., Jordan, M.I.: Unsupervised domain adaptation with residual transfer networks. In: Advances in Neural Information Processing Systems, pp. 136–144 (2016)
32. Long, M., Zhu, H., Wang, J., Jordan, M.I.: Deep transfer learning with joint adaptation networks. In: Proceedings of the 34th International Conference on Machine Learning, vol. 70, pp. 2208–2217 (2017). JMLR.org
33. Maaten, L.v.d., Hinton, G.: Visualizing data using t-SNE. J. Mach. Learn. Res. **9**, 2579–2605 (2008)
34. Netzer, Y., Wang, T., Coates, A., Bissacco, A., Wu, B., Ng, A.Y.: Reading digits in natural images with unsupervised feature learning. In: NIPS Workshop on Deep Learning and Unsupervised Feature Learning, vol. 2011, p. 5 (2011)
35. Nowozin, S., Cseke, B., Tomioka, R.: f-GAN: training generative neural samplers using variational divergence minimization. In: Advances in Neural Information Processing Systems, pp. 271–279 (2016)
36. Pan, S.J., Yang, Q.: A survey on transfer learning. IEEE Trans. Knowl. Data Eng. **22**(10), 1345–1359 (2010)
37. Russo, P., Carlucci, F.M., Tommasi, T., Caputo, B.: From source to target and back: symmetric bi-directional adaptive gan. In: Proceedings of the IEEE Conference on Computer Vision and Pattern Recognition, pp. 8099–8108 (2018)
38. Shen, J., Qu, Y., Zhang, W., Yu, Y.: Adversarial representation learning for domain adaptation. In: AAAI Conference (2018)
39. Shrivastava, A., Pfister, T., Tuzel, O., Susskind, J., Wang, W., Webb, R.: Learning from simulated and unsupervised images through adversarial training. In: The IEEE Conference on Computer Vision and Pattern Recognition (CVPR), vol. 3, p. 6 (2017)
40. Silberman, N., Hoiem, D., Kohli, P., Fergus, R.: Indoor segmentation and support inference from RGBD images. In: ECCV (2012)
41. Simonyan, K., Zisserman, A.: Very deep convolutional networks for large-scale image recognition. In: ICLR (2015)
42. Sun, B., Saenko, K.: Deep coral: Correlation alignment for deep domain adaptation. In: European Conference on Computer Vision, pp. 443–450. Springer, New York (2016)
43. Taigman, Y., Polyak, A., Wolf, L.: Unsupervised cross-domain image generation. In: 5th International Conference on Learning Representations, ICLR 2017, Toulon, April 24–26, 2017, Conference Track Proceedings (2017)
44. Tzeng, E., Hoffman, J., Zhang, N., Saenko, K., Darrell, T.: Deep domain confusion: maximizing for domain invariance (2014). Preprint. arXiv:1412.3474
45. Tzeng, E., Hoffman, J., Saenko, K., Darrell, T.: Adversarial discriminative domain adaptation. In: Proceedings of the IEEE Conference on Computer Vision and Pattern Recognition, pp. 7167–7176 (2017)
46. Venkateswara, H., Eusebio, J., Chakraborty, S., Panchanathan, S.: Deep hashing network for unsupervised domain adaptation. In: Proceedings of the IEEE Conference on Computer Vision and Pattern Recognition, pp. 5018–5027 (2017)
47. Villani, C.: Optimal Transport: Old and New, vol. 338. Springer Science & Business Media, Berlin/Heidelberg (2008)
48. Witten, I.H., Frank, E., Hall, M.A., Pal, C.J.: Data Mining: Practical Machine Learning Tools and Techniques. Morgan Kaufmann, Burlington, MA (2016)

49. Yan, H., Ding, Y., Li, P., Wang, Q., Xu, Y., Zuo, W.: Mind the class weight bias: weighted maximum mean discrepancy for unsupervised domain adaptation. In: Proceedings of the IEEE Conference on Computer Vision and Pattern Recognition, pp. 2272–2281 (2017)
50. Yu, Y., Szepesvári, C.: Analysis of kernel mean matching under covariate shift (2012). Preprint. arXiv:1206.4650
51. Zhang, K., Schölkopf, B., Muandet, K., Wang, Z.: Domain adaptation under target and conditional shift. In: International Conference on Machine Learning, pp. 819–827 (2013)
52. Zhang, W., Ouyang, W., Li, W., Xu, D.: Collaborative and adversarial network for unsupervised domain adaptation. In: Proceedings of the IEEE Conference on Computer Vision and Pattern Recognition, pp. 3801–3809 (2018)

Part III
Domain Alignment in the Image Space

Chapter 6
Unsupervised Domain Adaptation with Duplex Generative Adversarial Network

Lanqing Hu, Meina Kan, Shiguang Shan, and Xilin Chen

6.1 Introduction

Transfer learning is an effective technique which can reduce the labor and cost of collecting and labeling a large amount of related data to train a new model by borrowing knowledge from a different but sophisticated domain. Its difficulty lies in how to handle the domain shift between the related sophisticated domain (i.e., the source domain) and the testing domain (i.e., the target domain). This work focuses on a sub-problem of the general transfer learning, unsupervised domain adaptation, where the *labeled* source domain and the *unlabeled* target domain share the same task but lie in different distributions. The methods handling unsupervised domain adaptation can be categorized into two types: domain-invariant feature learning and domain transformation.

6.1.1 Domain-Invariant Feature Learning

In the earlier years, conventional subspace approaches like [8–10, 22, 27, 28] attempt to discover a common feature space of source and target domains where the domain gap is minimized as much as possible. For example, a representative method called geodesic flow kernel (GFK) in [9] is proposed to integrate infinite number of subspaces to model domain shift between the source and target domain, thus bridging the two domains. In [27] and [28], both source and target domain

L. Hu · M. Kan (✉) · S. Shan · X. Chen
Key Lab of Intelligent Information Processing of Chinese Academy of Sciences, Institute of Computing Technology, Chinese Academy of Sciences, Beijing, China
e-mail: lanqing.hu@vipl.ict.ac.cn

© Springer Nature Switzerland AG 2020
H. Venkateswara, S. Panchanathan (eds.), *Domain Adaptation in Computer Vision with Deep Learning*, https://doi.org/10.1007/978-3-030-45529-3_6

data are constrained to project into a low-rank common subspace to remove domain specific information for better adaptation.

Recently, the development of deep learning improves the fitting capability of models and achieves impressive progress on many tasks. Inspired by deep learning and metric learning [6], the methods in [17, 18] are proposed to extract deep domain-invariant features by minimizing the Maximum Mean Discrepancy (MMD) to reduce the domain gap. Apart from the straight-forward combination of deep networks and metric learning, there are also other methods which take advantage of image reconstruction to restrain the extracted feature to preserve more information of both source and target domains. In deep reconstruction-classification network (DRCN) [7], a latent representation, which can do classification task on labeled source domain and self reconstruction on unlabeled target domain simultaneously, is optimized to extract task-favorable information of both domains. In [2], a domain separation network (DSN) with self reconstruction and classification loss is designed to separate the domain specific representation and shared representation, leading to more accurate domain invariant feature. The newly proposed adversarial learning has shown its powerful ability in modeling the distribution discrepancy and thus better reducing the domain gap [4, 5, 30]. A typical method among these named Domain Adversarial Neural Networks (DANN) [5] proposes a gradient reversal layer which is added on the feature output layer to alleviate the discrepancy between source and target domain.

The methods above mainly focus on reducing the domain gap but without considering discriminality. In the latest years, several approaches start to concentrate on common discriminative feature learning [25, 26]. They mostly use pseudo-labeling to deal with domain discrepancy between source and target domain, which convert the unsupervised domain adaptation into semi-supervised one and thus achieve favorable performance improvement especially in the challenging scenarios.

6.1.2 Domain Transformation

There are also methods endeavoring to alleviate domain shift by domain transformation, e.g., transforming an image from target domain to source domain or vice versa. The method in [29] is a single directional transformation method, which transforms both source and target domains into source domain, thus aligning the target domain to the source. However, just single directional translation would result in a common feature biased to source domain. Thus, bidirectional transformation performs better in obtaining domain invariant feature. In [12], a Bi-shifting Auto Encoder (BAE) is proposed, which employs the sparse reconstruction to ensure the domain consistency during the process of bidirectional transformation. Furthermore, the methods in [15, 16, 19, 23] exploit the powerful GAN to implement bidirectional transformation via domain specific generators [15, 16] or weight sharing generator conditioned on a domain specific representation [19, 23]. Specially, CycleGAN [31] proposes a favorable cyclic domain translation. That is to say, a sample is firstly transformed to another domain, and then the transformed sample is further transformed back to the original domain. This kind of cyclic constraint effectively enforces the generated images to have the specific domain style.

In these existing methods, the domain invariant feature learning and domain transformation methods usually do not consider the discriminality, i.e., the category of a sample may be changed during domain adaptation. There are several recent works utilizing pseudo-labeling for improving the discriminality of the extracted features [25, 26] achieve further improvement on domain adaptation, however, they consider less about the domain invariant constraint. In this work, we also consider both domain invariant and discriminative feature learning for unsupervised domain adaptation. Specifically, a novel duplex generative adversarial net for bidirectional domain transformation named DupGAN is proposed to ensure the latent feature representation domain invariant. Besides, each of the duplex discriminators is designed for reality/falsity discrimination as well as category classification, to enforce the latent feature representation discriminative. An overview is shown in Fig. 6.1.

Briefly, the contributions of this work lie in three folds: (1) A novel generative adversarial network with duplex discriminators named DupGAN is proposed to extract domain invariant representations via bidirectional domain transformation; (2) The duplex discriminators with both reality/falsity discrimination and image category classification ensure latent representation discriminative; (3) DupGAN outperforms the state-of-the-art methods on unsupervised domain adaptation of digit classification and object recognition. This chapter is partially based on earlier work from [11].

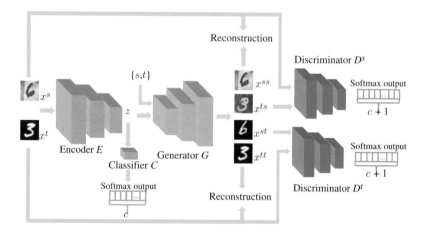

Fig. 6.1 Overall structure of our Duplex Generative Adversarial Net (DupGAN). The encoder E (orange cubes), the generator G (green cubes) conditioned on the domain code $\{s, t\}$ and the duplex discriminators D^s and D^t (blue cubes) collaborate with each other to obtain the domain-invariant as well as discriminative feature z

6.2   Method

In unsupervised domain adaptation for classification tasks, source domain is labeled while target domain is unlabeled, with the data denoted as $X^s = \{(x_i^s, y_i^s)\}_{i=1}^n$ and $X^t = \{x_j^t\}_{j=1}^m$, respectively. As introduced above, the source and target domains share c identical categories but follow different data distribution. The goal is to learn a model for classification on target domain with both labeled training data $\{x_i^s, y_i^s\}$ of source domain and unlabeled training data $\{x_j^t\}$ of target domain. In other words, knowledge of source domain is transferred to target domain for better classification on the testing set of target domain.

To achieve satisfactory performance on target domain in unsupervised domain adaptation, the key is to discover the common knowledge from source domain to benefit the target domain. Aiming for this, our method proposes a novel GAN with duplex discriminators named DupGAN to achieve domain invariant and discriminative feature for classification on target domain. As illustrated in Fig. 6.1, our proposed DupGAN mainly consists of three successive modules. Specifically, the **encoder**, denoted as E, projects input images X from either domain into a shared representation space $Z = E(X)$, which is the ultimate part for testing with the objective to be domain invariant and discriminative. To fulfill the goal, a generator followed by duplex discriminators are elaborately designed. The **generator**, denoted as G, decodes a latent representation $z \in Z$ into a source/target domain image conditioned on a domain indicator a. The **duplex discriminators** D^s (for source domain) and D^t (for target domain) pitted against the generator not only distinguish the reality/falsity of the input images but also classify the real images. Then those generated images from the generator that can cheat the discriminators will look real as well as preserve the category information. This further enforces the latent representation fed into the generator to be domain invariant and discriminative. Briefly, the rationality behind this method is that the gradient direction for z from duplex discriminators will be compromised when they conflict with each other, leading to a common subspace, especially when the commonality is dominant which is also the basis of domain adaptation as proved in [1].

6.2.1   Encoder and Classifier

The encoder is to project an image from either source domain or target domain into a shared latent representation, formulated as follows:

$$z^s = E(x^s), x^s \in X^s,$$
$$z^t = E(x^t), x^t \in X^t, \tag{6.1}$$
$$z = \{z^s, z^t\} \in Z,$$

where E is the encoder with parameter denoted as W^E, z^s is the encoded feature of source domain and z^t is the encoded feature of target domain.

The latent representation z is expected to be domain invariant and discriminative, which is guaranteed by the bidirectional domain transformation from generator and the adversarial learning between generator and duplex discriminator. The design of generator and duplex discriminators will be detailed in Sects. 6.2.2 and 6.2.3, respectively.

For the categorial classification task, a classifier C is constructed on the latent representation z, which is formulated as:

$$
\begin{aligned}
c^s &= C(E(x^s)), x^s \in X^s, \\
c^t &= C(E(x^t)), x^t \in X^t,
\end{aligned}
\tag{6.2}
$$

where C stands for the classifier with parameter denoted as W^C. And the objective function of this classifier is the plain softmax function:

$$
L_c = H(c^s, y^s) + H(c^t, y^t),
\tag{6.3}
$$

where $H(\cdot, \cdot)$ is the cross entropy loss. The y^s and y^t are the ground-truth class label of source domain samples and estimated class labels of target domain samples, defined as below:

$$
\begin{aligned}
y^s &= [\underbrace{0, 0, \cdots, 0}_{i-1}, 1, \underbrace{0, \cdots, 0}_{c-i}], cat(x^s) = i, \\
y^t &= [\underbrace{0, 0, \cdots, 0}_{j-1}, 1, \underbrace{0, \cdots, 0}_{c-j}], cat(x^t) = j,
\end{aligned}
\tag{6.4}
$$

where $cat(\cdot)$ means the category of the input sample.

The classifier does the final classification for testing with the domain-invariant features as input. Besides, it can also improve the discriminality of the latent representation z during the training stage.

What deserves to be detailed is that the categorial label y^s of the source domain image is available and can be used directly, while the label y^t for target domain is not. Therefore, y^t is obtained from the prediction of classifier C via pseudo labeling. More specifically, at the beginning, the category information of the whole target domain is unavailable, so the classifier is firstly pretrained with only the source domain. Then according to the dominance of commonality in [1], a few target domain samples closer to source domain are selected with pseudo labels for further training the classifier and the discriminators. As the iteration goes on, more target samples are added in to train more accurate DupGAN.

6.2.2 Generator

The generator is designed to decode a latent representation z to both source and target domains with a conditional domain indicator. The intention behind this is that if a latent representation of either domain can be well decoded into both source and target domains, it should be a common representation between domains, otherwise it can only be well decoded to one domain.

Specifically, the generator G takes the concatenation of a representation z and a domain indicator $a \in \{s, t\}$ as input, and then decodes it into source and target domains. Here, when $a = s$ specifying source domain, the generator attempts to generate an image with source domain style. When $a = t$, the generator tries to output an image with target domain style. The domain transformations by the generator are formulated as follows:

$$x^{ss} = G(z^s, s) = G(E(x^s), s),\tag{6.5}$$

$$x^{tt} = G(z^t, t) = G(E(x^t), t),\tag{6.6}$$

$$x^{st} = G(z^s, t) = G(E(x^s), t),\tag{6.7}$$

$$x^{ts} = G(z^t, s) = G(E(x^t), s).\tag{6.8}$$

Equations (6.5) and (6.6) are the self-reconstruction generation, i.e., transforming a source domain sample to source domain, or a target domain sample to target domain. As a general rule, when an image is transformed into its original domain, the output should look the same as itself. So, a reconstruction loss for the generator is exploited to keep the image unchanged after self-domain transformation, which is formulated as:

$$L_r = ||x^{ss} - x^s||_2^2 + ||x^{tt} - x^t||_2^2.\tag{6.9}$$

Equations (6.7) and (6.8) represent the cross domain transformation, i.e., from source domain to target domain or vice versa. In this case, the generated image x^{st} should look like a real target domain image and also its category should be consist with that of x^s. Similarly, the generated image x^{ts} should look like a real source domain image, and also its category should be consist with that of x^t. These are ensured by the adversarial learning between the generator and the following duplex discriminators, i.e., source and target domain discriminators. Specifically, given the well optimized duplex discriminators, x^{st} (Eq. (6.7)) and x^{ts} (Eq. (6.8)) are optimized to be classified as real samples and correct categories by the discriminators. More details are provided in Eq. (6.15) in Sect. 6.2.3.

Guided by the reconstruction constraint and domain adversarial objective for Eqs. (6.5)–(6.8), a generator for bidirectional domain transformation can be obtained, which further enforces the input of the generator, i.e., the latent representation z, to be domain invariant.

6.2.3 Duplex Discriminators

Like the conventional GAN, the duplex discriminators are designed as the adversary of the generator. Specifically, the source domain discriminator D^s is to distinguish the real source domain samples x^s from the generated source-style samples x^{ts}. The target domain discriminator D^t is to distinguish the real target domain samples x^t from the generated target-style samples x^{st}. Besides distinguishing the real and fake samples, the discriminators also classify the category of the real sample in order that the image content especially the category information can be well preserved. To do category classification, the discriminators are augmented to have $c+1$ output nodes, where the first c nodes are for classification of real images, one node for each category, and the last node represents the fake images.

The adversarial training between duplex discriminators and generator together with encoder is in alternative manner. When training the duplex discriminators, the generator and encoder is fixed and the objective is to distinguish the generated images from real images of both domains, and also to correctly classify the real images. This is specified as follows:

$$L_d = H\left(D^s(x^s), \tilde{y}^s\right) + H\left(D^s(x^{ts}), \tilde{y}^{ts}\right) + H\left(D^t(x^t), \tilde{y}^t\right) + H\left(D^t(x^{st}), \tilde{y}^{st}\right),$$
$$(6.10)$$

where D^s and D^t are the source and target domain discriminators with parameters denoted as W^D. The first two terms are for the discriminator of source domain, while the last two terms are for the discriminator of target domain. Among the images input into the discriminator D^s, the real source-domain image x^s should be classified into its ground truth category, while the generated source-style image x^{ts} should be classified to be fake rather than any real category. So, in Eq. (6.10), the corresponding labels for x^s and x^{ts} are as below:

$$\tilde{y}^s = [y^s, 0] = [\overbrace{0, 0, \cdots, 0}^{i-1}, 1, \underbrace{0, \cdots, 0}_{c-i}, 0], \, cat(x^s) = i, \qquad (6.11)$$

with y^s spanning the first part.

$$\tilde{y}^{ts} = [\underbrace{0, 0, \cdots, 0}_{c}, 1], \forall x^{ts}, \qquad (6.12)$$

where $\hat{y}^t$ is a concatenation of y^t in Eq. (6.4) and a scalar 0. The y^s represents the pseudo category label of x^s, while the last node in $\tilde{y}^s$, i.e., 0, is for the falsity. The $\tilde{y}^{st}$ is a concatenation of a c-dim vector $\mathbf{0}$ and a scalar 1, meaning that x^{st} is classified as the fake class.

Similarly, among the images input into the discriminator D^t, the real target-domain image x^t should be classified to its category estimated from classifier C (Eq. (6.4)), while the generated target-style image x^{st} should be classified to be fake rather than any real category. So, in Eq. (6.10), the corresponding labels for x^t and x^{st} are as follows:

$$\hat{y}^t = [y^t, 0] = [\overbrace{\underbrace{0, 0, \cdots, 0}_{j-1}, 1, \underbrace{0, \cdots, 0}_{c-j}}^{y^t}, 0], cat(x^t) = j, \tag{6.13}$$

$$\tilde{y}^{st} = [\underbrace{0, 0, \cdots, 0}_{c}, 1], \forall x^{st}, \tag{6.14}$$

where $\hat{y}^t$ is a concatenation of y^t in Eq. (6.4) and a scalar 0. The y^t represents the pseudo category label of x^t, while the last node in $\tilde{y}^t$, i.e., 0, is for the falsity. The $\tilde{y}^{st}$ is a concatenation of a c-dim vector $\mathbf{0}$ and a scalar 1, meaning that x^{st} is classified as the fake class.

After optimizing the duplex discriminators, they are fixed to further guide the generator and encoder with conflict objective with the discriminators. Specifically, a good generator should synthesize images that can cheat the discriminator to be classified as real ones, with the objective formulated as follows:

$$L_g = H\left(D^t(x^{st}), \tilde{y}^{st}\right) + H\left(D^s(x^{ts}), \tilde{y}^{ts}\right), \tag{6.15}$$

where the first term means that the transformed image x^{st} from the generator should be classified into one of c real target classes via D^t, to make it look realistic and also preserve the category information. So, for x^{st}, the corresponding label $\tilde{y}^{st}$ is set as:

$$\tilde{y}^{st} = [y^s, 0] = [\overbrace{\underbrace{0, 0, \cdots, 0}_{i-1}, 1, \underbrace{0, \cdots, 0}_{c-i}}^{y^s}, 0], cat(x^s) = i. \tag{6.16}$$

Similarly, the second term in Eq. (6.15) means that the transformed image x^{ts} from the generator should be classified into one of c real source classes via D^s, with the corresponding labels $\tilde{y}^{st}$ set as:

$$\tilde{y}^{ts} = [y^t, 0] = [\overbrace{\underbrace{0, 0, \cdots, 0}_{j-1}, 1, \underbrace{0, \cdots, 0}_{c-j}}^{y^t}, 0], cat(x^t) = j. \tag{6.17}$$

As the target domain samples are unlabeled, the pseudo class label y^t used here is the estimated pseudo label from classifier C in Eq. (6.4).

Taking all the above into consideration, Eqs. (6.10) and (6.15) form the adversarial training between the generator and duplex discriminators, which are alternatively optimized. With the constraint of bidirectional translation and duplex discriminators, the generated samples x^{ss} and x^{ts} will look like the real source domain samples with category preserved, while x^{tt} and x^{st} will look like the real target samples as well as keep the category after translation. To satisfy the objective of reality and category preservation during translation, the latent representation z is enforced to be domain invariant and discriminative, otherwise it can only be well transformed into one domain, or with category distortion. The overall training process is detailed in Sect. 6.2.4.

6.2.4 Overall Objective

The overall training process is an alternative procedure. One training stage is to train the duplex discriminators and the other is to train the encoder, generator and classifier. Specifically, with the encoder, generator, and classifier fixed, the training objective function for duplex discriminators D^s and D^t is as follows:

$$\min_{W^D} L_D = \sum_{x^s \in X^s, x^t \in X^t} L_d. \tag{6.18}$$

With the duplex discriminators fixed, the training objective function for the encoder E, classifier C and generator G is:

$$\min_{W^E, W^C, W^G} L_{ECG} = \sum_{x^s \in X^s, x^t \in X^t} \left(L_g + \alpha L_r + \beta L_c \right), \tag{6.19}$$

where α and β are balance parameters.

The above two objectives can be easily optimized via stochastic gradient descent as most existing works do. The detailed optimization process is displayed in Algorithm 1.

6.3 Experimental Settings

All experiments are conducted on two standard unsupervised domain adaptation benchmarks of digit and object classification. The datasets and experimental settings are detailed in Sect. 6.3.1. To investigate the effectiveness of our proposed method, it is compared with a few state-of-the-art methods, including DANN [4, 5] and ADDA [30], classification and reconstruction combining methods DSN [2] and DRCN [7], GAN-based methods CoGAN [15] and UNIT [16], clustering and pseudo-labeling methods kNN-Ad [26] and ATDA [25].

Algorithm 1 Optimization procedure of DupGAN

Input: The source domain sample X^s and category label Y^s, target domain sample X^t.
Output: The parameters of whole network, $W = \{W^E, W^C, W^G, W^D\}$.
 1: Pre-train E and C with images in X^s.
 2: Pseudo-labeling those target domain samples with high confidence from C.
 3: **while** not converge **do**
 4:      Update W^D by minimizing L_D in Eq. (6.18) through the gradient descent:
        $W^D \leftarrow W^D - \eta \frac{\partial L_D}{\partial W^D}$
 5:      Update W^G, W^C and W^E by minimizing L_{ECG} in Eq. (6.19) through the gradient descent:
        $W^G \leftarrow W^G - \eta \frac{\partial L_g + \alpha \partial L_r}{\partial W^G}$
        $W^C \leftarrow W^C - \eta \beta \frac{\partial L_c}{\partial W^C}$
        $W^E \leftarrow W^E - \eta (\beta \frac{\partial L_c}{\partial W^C} \times \frac{\partial W^C}{\partial W^E} + \frac{\partial L_g}{\partial W^G} \times \frac{\partial W^G}{\partial W^E} + \alpha \frac{\partial L_r}{\partial W^G} \times \frac{\partial W^G}{\partial W^E})$
 6:      Update the pseudo labels of those target domain samples with high confidence from C.
 7: **end while**

6.3.1 Datasets and Settings

For **digit classification**, the datasets of MNIST [14], MNIST-M [4, 5], USPS [3], SYN DIGITS [4, 5] and SVHN [20] are used for evaluating all the methods. All five datasets contain images of digits 0~9 but with different styles. MNIST is a handwritten digits dataset composed of 60, 000 training and 10, 000 testing images. USPS is a US postal handwritten digits dataset, consisting of 7291 training and 2007 testing images. The Street View House Number dataset SVHN contains 73, 257 training images, 26, 032 testing images and 531, 131 extra training images. MNIST-M and SYN DIGITS are both synthesized digit images. MNIST-M is constructed by merging the clip of the background from BSDS500 datasets [21] to MNIST, consisting of 60, 000 training and 10, 000 testing images, same as those of MNIST. SYN DIGITS consists of 500, 000 images generated from Windows fonts by varying the text, positioning, orientation, background and colors. Exemplar images and experimental settings are summarized in Fig. 6.2 and Table 6.1. Following the existing works, five evaluations based on these datasets are reported for unsupervised domain adaptation of digit classification.

MNIST ↔ USPS In MNIST → USPS, 60, 000 labeled training images from MNIST and 7291 unlabeled training images from USPS are used as the training set of unsupervised domain adaptation. Then the evaluation is deployed on the 2007 testing images in USPS. While in USPS → MNIST, 7291 labeled training images from USPS and 60, 000 unlabeled training images from MNIST are used as the training set of unsupervised domain adaptation. Then the evaluation is deployed on the 10, 000 testing images in MNIST. As seen from Fig. 6.2, the visual discrepancy between MNIST and USPS is not so large, so the adaptation between MNIST and USPS is easier.

SVHN ↔ MNIST In SVHN → MNIST, 73,257 labeled training images from SVHN and 60,000 unlabeled training images from MNIST are used as the training

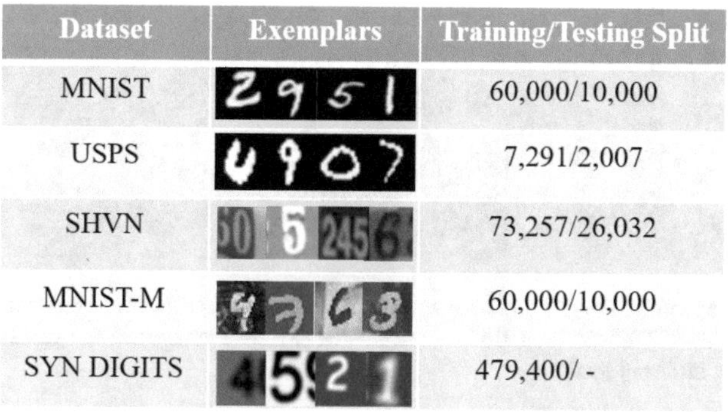

Dataset	Exemplars	Training/Testing Split
MNIST		60,000/10,000
USPS		7,291/2,007
SHVN		73,257/26,032
MNIST-M		60,000/10,000
SYN DIGITS		479,400/ -

Fig. 6.2 The exemplar images and image numbers of digits datasets, including MNIST, USPS, SVHN, MNIST-M and SYN DIGITS. All these datasets contain the same 10 categories of digits, i.e., $\{0, 1, \cdots, 9\}$ but with different domain style, e.g., appearance and background

Table 6.1 The detailed experimental settings for digit classification and object recognition

Experiment		#Source training	#Target training	#Testing
Digit Cls	MNIST → USPS	60,000	7291	2007
	USPS → MNIST	7291	60,000	10,000
	SVHN → MNIST	73,257	60,000	10,000
	SVHN^{extra} → MNIST	531,131	60,000	10,000
	MNIST → SVHN	60,000	73,257	26,032
	MNIST → MNIST-M	60,000	60,000	10,000
	SYN DIGITS → SVHN	479,400	73,257	26,032
Object Rec	Amazon (A) → Webcam (W)	2817	795	795
	Webcam (W) → Amazon (A)	795	2817	2817
	Amazon (A) → DSLR (D)	2817	498	498
	DSLR (D) → Amazon (A)	498	2817	2817

set of unsupervised domain adaptation. The 10,000 testing images of MNIST are exploited as the testing set. While in MNIST → SVHN, 60,000 labeled training images from MNIST and 73,257 unlabeled training images from SVHN are used as the training set. The 26,032 testing images from SVHN are exploited as the testing data. Note that there is a third experiment dubbed as SVHN^{extra} → MNIST where all labeled extra training set from SVHN (SVHN^{extra}) containing 531,131 images is used as the source domain training set. As seen from Fig. 6.2, SVHN contains more variations than MNIST and thus is more informative. So SVHN → MNIST is easier than MNIST → SVHN. What deserves to be mentioned is that MNIST → SVHN is the most challenging experiment in unsupervised domain adaptation of digit classification.

MNIST → MNIST-M MNIST-M is a synthesized dataset generated from MNIST digits but with more variations than MNIST. It is also a challenging task. In this experiment, 60,000 labeled training images from MNIST and 60,000 unlabeled

Dataset	Exemplars	Training/Testing Split
Amazon		2,817/2,817
Webcam		795/795
DSLR		498/498

Fig. 6.3 The exemplar images and image numbers of object classification datasets, i.e., Office-31 including Amazon, Webcam and DSLR, composing of office images in the same 31 classes but collected from different media

training images from MNIST-M are used as the training set of unsupervised domain adaptation. And the testing split consisting of 10,000 images from MNIST-M are utilized as the testing set.

SYN DIGITS → SVHN 479,400 labeled digit images from the synthesized digit dataset SYN DIGITS are used as source domain training data, and 73,257 unlabeled images from SVHN are used as the target domain training set. Then the testing split of SVHN with 26,032 images is utilized for testing. Note that in this evaluation, the gray images perform similarly as the RGB images, so gray images are used for faster training.

For **object recognition**, commonly used benchmark Office-31 [24] are exploited. This dataset contains 4110 images of 31 categories collected from three distinct domains: Amazon (A) consisting of 2817 images from Amazon online shopping Website, Webcam (W) containing 795 pictures captured by the webcam monitor and DSLR (D) composed of 498 images photographed by the digital single lens reflex camera, detailed in Fig. 6.3 and Table 6.1. On this dataset, we mainly evaluate on the challenging sub-experiments of $A \leftrightarrow W$ and $A \leftrightarrow D$. Following the standard unsupervised domain adaptation training protocol in [7], for each evaluation, all labeled source data and unlabeled target data are used for training. At the same time, all unlabeled target data are also used for testing.

6.3.2 Implementation Details

For all the experiments, the detailed network architectures of our method on digits experiments are shown in Tables 6.2 and 6.3, which is the same as the state-of-the-arts [16, 25]. The encoder architecture of our method on Office-31 experiments is AlexNet [13], which is the same as the compared works like [7], guaranteeing fair comparison. In addition, for training convenience, all the image pixel values are linearly rescaled from [0, 255] to [−1.0, 1.0]. Correspondingly, TanH function is chosen as the output activation function of the generator, which can rescale all inputs

Table 6.2 The architectures of our DupGAN used in MNIST → MNIST-M and MNIST ↔ USPS

	Layer	Architecture of MNIST → MNIST-M	Architecture of MNIST ↔ USPS
Encoder	1	CONV-(C32,K5,S1), ReLU, MP-(K2,S2)	CONV-(C32,K5,S1), PReLU, MP-(K2,S2)
	2	CONV-(C48,K5,S1), BN, ReLU, MP-(K2,S2)	CONV-(C64,K5,S1), PReLU, MP-(K2,S2)
	3	CONV-(C100,K8,S1), BN, ReLU	CONV-(C128,K7,S1), PReLU
	4	–	CONV-(C256,K1,S1), PReLU
Classifier	1	CONV-(C100,K1,S1), BN, ReLU	FC-(C10)
	2	FC-(C10)	
Generator	1	DECONV-(C256,K4,S4), BN, LReLU	DECONV-(C128,K4,S4), BN, PReLU
	2	DECONV-(C128,K4,S2), BN, LReLU	DECONV-(C64,K3,S2), BN, PReLU
	3	DECONV-(C64,K4,S2), BN, LReLU	DECONV-(C32,K3,S2), BN, PReLU
	4	DECONV-(C3,K4,S2), TanH	DECONV-(C1,K6,S1), TanH
Discriminators	1	CONV-(C32,K5,S1), ReLU, MP-(K2,S2)	CONV-(C32,K5,S1), PReLU, MP-(K2,S2)
	2	CONV-(C48,K5,S1), BN, ReLU, MP-(K2,S2)	CONV-(C64,K5,S1), PReLU, MP-(K2,S2)
	3	CONV-(C100,K8,S1), BN, ReLU	CONV-(C512,K4,S1), PReLU
	4	CONV-(C100,K1,S1), BN, ReLU	FC-(C11)
	5	FC-(C11)	

In each layer, the (C, K, S) stand for number of output channels, kernel size, and stride, respectively. The first 3 layers of the encoders and discriminators of MNIST → MNIST-M share weights for two domains, respectively

into $[-1.0, 1.0]$. As for the balance parameters α and β in the objective function Eq. (6.19), in the experiments of MNIST ↔ USPS, SYN DIGITS → SVHN as well as SVHN → MNIST, they are empirically set as 10.0 and 1.0. While in MNIST → SVHN, MNIST → MNIST-M and Office-31, they are set as 1.0 and 1.0.

6.4 Experiments

6.4.1 Unsupervised Domain Adaptation on Digit Classification

For digit classification, the unsupervised domain adaptation experiments are conducted on MNIST ↔ USPS, SVHN ↔ MNIST, MNIST→ MNIST-M and SYN

Table 6.3 The architectures of our DupGAN used in SYN DIGITS→ SVHN and SVHN ↔ MNIST

	Layer	Architecture of SYN DIGITS→ SVHN & SVHN → MNIST	Architecture of MNIST → SVHN
Encoder	1	CONV-(C64,K5,S2), BN, LReLU	CONV-(C64,K5,S1), ReLU, MP-(K3,S2)
	2	CONV-(C128,K5,S2), BN, LReLU	CONV-(C64,K5,S1), ReLU, MP-(K3,S2)
	3	CONV-(C256,K5,S2), BN, LReLU	CONV-(C128,K5,S1), ReLU
	4	CONV-(C512,K4,S1), BN, LReLU	CONV-(C512,K4,S1), ReLU
Classifier	1	FC-(C10)	CONV-(C3072,K4,S1), BN, ReLU
	2		CONV-(C2048,K1,S1), ReLU
	3		FC-(C10)
Generator	1	DECONV-(C256,K4,S4), BN, LReLU	DECONV-(C256,K4,S4), BN, PReLU
	2	DECONV-(C128,K4,S2), BN, LReLU	DECONV-(C128,K4,S2), BN, PReLU
	3	DECONV-(C64,K4,S2), BN, LReLU	DECONV-(C64,K4,S2), BN, PReLU
	4	DECONV-(C3,K4,S2), TanH	DECONV-(C3,K4,S2), TanH
Discriminators	1	CONV-(C64,K5,S1), BN, LReLU	CONV-(C64,K5,S1), ReLU, MP-(K3,S2)
	2	CONV-(C128,K5,S1), BN, LReLU	CONV-(C64,K5,S1), ReLU, MP-(K3,S2)
	3	CONV-(C256,K5,S1), BN, LReLU	CONV-(C128,K5,S1), ReLU
	4	CONV-(C512,K4,S1), BN, LReLU	CONV-(C3072,K4,S1), BN, ReLU
	5	FC-(C11)	CONV-(C2048,K1,S1), ReLU
	6		FC-(C11)

In each layer, the (C, K, S) stand for number of output channels, kernel size, and stride, respectively. The first 4 layers of the encoders and discriminators of MNIST → SVHN share weights for two domains, respectively

DIGITS → SVHN. For fair comparison, we follow the same training and testing protocol with the compared methods [4, 7, 16, 25].

The performance of all methods are shown in Table 6.4. As can be seen, compared with the vanilla adversarial learning method DANN [4, 5] and ADDA [30], CoGAN [15] and UNIT [16] gain more performance improvement by further utilizing the image generation and adversarial learning for information preservation. The methods kNN-Ad [26] and ATDA [25] perform the best among these compared methods on the challenging experiments SVHN ↔ MNIST mainly because they mine more discriminative information. Our proposed DupGAN outperforms all the

Table 6.4 The results of unsupervised domain adaptation of digit classification

Method	MNIST → USPS	USPS → MNIST	SVHN → MNIST	MNIST → SVHN	SVHN^{extra} → MNIST	MNIST → MNIST-M	SYN DIGITS → SVHN	Avg
Source Only	86.75	75.52	62.19	33.70	73.67	57.10	85.50	67.78
DANN [4, 5]	85.10	73.00	73.85	–	–	81.50	90.30	–
DRCN [7]	91.80	73.67	81.97	40.05	–	–	–	–
ADDA [30]	92.87*	93.75*	76.00	–	86.37	–	–	–
DSN [2]	91.30*	73.20*	82.70	–	–	83.20	91.20	–
CoGAN [15]	95.65	93.15	–	–	–	–	–	–
UNIT [16]	95.97	93.58	–	–	90.53	–	–	–
kNN-Ad [26]	–	–	78.80	40.30	–	86.70	–	–
ATDA [25]	93.17*	84.14*	85.80	52.80	91.45*	94.00	92.90	84.89
DupGAN (Ours)	**96.01**	**98.75**	**92.46**	**62.65**	**96.42**	**94.09**	**93.08**	**90.49**
Target Only	95.02	98.96	98.97	87.74	98.97	95.57	90.97	95.17

The test protocols of all the compared methods including ours are the same, so the results of other methods are directly copied from the original papers except our reproduced results marked with stars which is missing in the original evaluations related methods, especially gaining an improvement up to 10% on the challenging MNIST → SVHN, demonstrating that our method obtains more domain-invariant as well as discriminative feature representation. This can be safely attributed to our effective design of bidirectional domain transformation via duplex path of adversarial optimization between the generation and discrimination.

What deserves to be mentioned is that in MNIST → USPS and SYN DIGITS → SVHN, DupGAN even outperforms the "Target Only" model which roughly serves as the groundtruth model. This may be because that the source domain contains more variations or information than target domain that are beneficial for target domain, which is exactly the case where domain adaptation is applicable.

To verify the effect of our method on alleviating distribution gap, the feature space before and after adaptation on SVHN → MNIST and SYN DIGITS → SVHN are visualized in Fig. 6.4. As can be seen, the discrepancy is significantly reduced and the obtained feature is quite discriminative in the common representation space of our proposed DupGAN compared with the original one, demonstrating the capability of our DupGAN to migrate distributions with category structure well preserved. Especially on SYN DIGITS → SVHN, even if the two domains are originally close to each other, our DupGAN can further alleviate the distribution gap and improve the discriminality of the latent feature space.

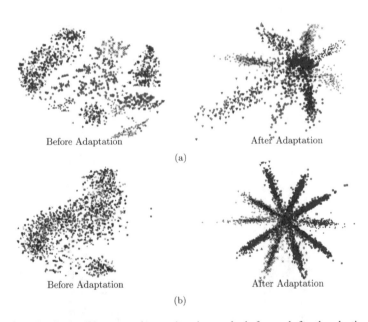

Before Adaptation   After Adaptation

(a)

Before Adaptation   After Adaptation

(b)

Fig. 6.4 The distribution of source and target domain samples before and after the adaption on two different digit classification experiments (**a**) SVHN → MNIST and (**b**) SYN DIGITS → SVHN. The source and target domain samples are shown in red and blue respectively, and the categories are drawn with different shapes

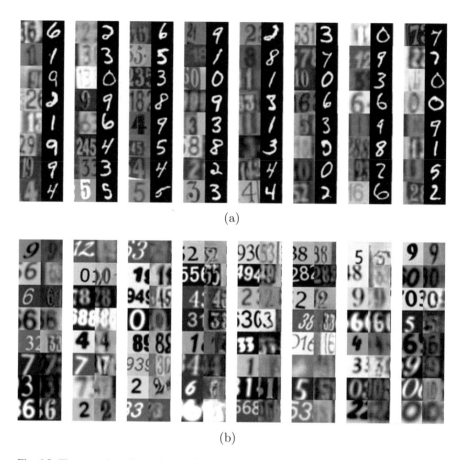

(a)

(b)

Fig. 6.5 The exemplars of domain transformation on two experiments (**a**) from SVHN to MNIST and (**b**) from SYN DIGITS to SVHN. In every two columns, the left and right images are the images from real source domain and their corresponding transformed images with target domain style respectively. As can be seen, the transformed images look like real target domain samples with category well preserved. Note that the digits in the center of images in SVHN and SYN DIGITS represent the major content

Furthermore, we visualize the transformed images of DupGAN to verify the effectiveness of duplex transformation. The translated images of SVHN → MNIST and SYN DIGITS → SVHN are shown in Fig. 6.5. As can be seen, the source domain images are well transformed into target domain with realistic target style. Moreover, only few images have the category distortion, indicating that DupGAN can well preserve the category information between the bidirectional transformation.

6.4.2   Unsupervised Domain Adaptation on Object Recognition

For more intensive investigation, we also conduct unsupervised domain adaptation experiments on Office-31 for object classification task. As we have conducted comprehensive comparison on the digits datasets, we just supplement the comparison with those typical methods on Office-31, including the classic adversarial learning method DANN [5] and classification-reconstruction method DRCN [7]. As shown in Table 6.5, DupGAN still performs the best among all compared methods with significant improvement, about 5.3% on average. This further verifies the effectiveness and robustness of our proposed DupGAN on different tasks.

6.4.3   Ablation Study

To investigate the effect of each separable part of the proposed DupGAN, the ablation study on MNIST $\rightarrow$ USPS and SVHN $\rightarrow$ MNIST is conducted. The detailed results are shown in Table 6.6. Specifically, to explore the effectiveness of discriminator with category preservation, duplex discriminators, and the generator, we evaluate the model removing the reality/falsity node of discriminators, referred to as "DupGAN-woA"; the model completely erasing the discriminators, i.e., only with classifier and generator with self reconstruction, referred to as "DupGAN-woAD"; and the model without the generator and discriminators, i.e., with only the classifier C and the pseudo-labeling process referred to as "DupGAN-woADG".

As can be seen, in both experiments, the performance of our DupGAN degenerates if without any separable part, which illustrates the indispensability of all parts for the whole framework. More significantly, on the challenging adaptation experiment of SVHN $\rightarrow$ MNIST, "DupGAN-woAD" outperforms "DupGAN-

Table 6.5 The results of unsupervised domain adaptation on Office-31 for object classification

Method	A $\rightarrow$ W	W $\rightarrow$ A	A $\rightarrow$ D	D $\rightarrow$ A	Avg
Source only	61.6	49.8	63.8	51.1	56.6
DAN [17]	68.5	53.1	67.0	54.0	60.7
DANN [4, 5]	72.6	52.7	67.1	54.5	61.7
DRCN [7]	68.7	54.9	66.8	56.0	61.6
DupGAN (Ours)	**73.2**	**59.1**	**74.1**	**61.5**	**67.0**

The test protocols of all the compared methods including ours are the same, so the results of other methods are directly copied from the original papers

Table 6.6 Ablation study on digit classification of MNIST $\rightarrow$ USPS and SVHN $\rightarrow$ MNIST

	DupGAN-woADG	DupGAN-woAD	DupGAN-woA	DupGAN
MNIST $\rightarrow$ USPS	93.32	93.82	94.57	**96.01**
SVHN $\rightarrow$ MNIST	60.18	67.43	68.30	**92.46**

woADG" with 7% improvement, indicating the effectiveness of the reconstruction constraint of generator. Then the performance of "DupGAN-woAD" is surpassed by "DupGAN-woA" which demonstrates the importance of category preservation of duplex discriminators. The complete DupGAN performs the best with a large margin improvement (24%) over "DupGAN-woA" showing that the adversarial learning plays an essential role in the whole framework. Overall speaking, both adversarial learning and category information preservation during domain transformation in our method contribute a lot to unsupervised domain adaptation problem.

6.5 Conclusion

In this work, we propose a novel GAN with duplex discriminators dubbed as DupGAN for unsupervised domain adaptation. Our DupGAN is composed of three essential parts, i.e., an encoder extracting domain-invariant feature representation, a generator decoding the representation to source and target domains, and the duplex discriminators ensuring the reality and preserving the category information. The whole framework works as a harmonious and self-contained system to ensure the latent representation to be domain invariant and discriminative. The state-of-the-art performance on several unsupervised domain adaptation benchmarks show the effectiveness of our proposed DupGAN.

References

1. Ben-David, S., Blitzer, J., Crammer, K., et al.: A theory of learning from different domains. Mach. Learn. **79**(1–2), 151–175 (2010)
2. Bousmalis, K., Trigeorgis, G., Silberman, N., Krishnan, D., Erhan, D.: Domain separation networks. In: Advances in Neural Information Processing Systems (NIPS), pp. 343–351 (2016)
3. Denker, J.S., Gardner, W.R., Graf, H.P., Henderson, D., Howard, R.E., Hubbard, W.E., Jackel, L.D., Baird, H.S., Guyon, I.: Neural network recognizer for hand-written zip code digits. In: Advances in Neural Information Processing Systems (NIPS), pp. 323–331 (1988)
4. Ganin, Y., Lempitsky, V.: Unsupervised domain adaptation by backpropagation. In: International Conference on Machine Learning (ICML) (2015)
5. Ganin, Y., Ustinova, E., Ajakan, H., Germain, P., et al.: Domain-adversarial training of neural networks. IEEE J. Mach. Learn. Res. **17**(1), 2096–2030 (2016)
6. Geng, B., Tao, D., Xu, C.: DAML: domain adaptation metric learning. IEEE Trans. Image Process. **20**(10), 2980–2989 (2011)
7. Ghifary, M., Kleijn, W.B., Zhang, M., Balduzzi, D., Li, W.: Deep reconstruction-classification networks for unsupervised domain adaptation. In: European Conference on Computer Vision (ECCV), pp. 597–613 (2016)
8. Gong, B., Grauman, K., Sha, F.: Connecting the dots with landmarks: discriminatively learning domain-invariant features for unsupervised domain adaptation. In: International Conference on Machine Learning (ICML), pp. 222–230 (2013)
9. Gong, B., Shi, Y., Sha, F., Grauman, K.: Geodesic flow kernel for unsupervised domain adaptation. In: IEEE Conference on Computer vision and Pattern Recognition (CVPR), pp. 2066–2073 (2012)

10. Gopalan, R., Li, R., Chellappa, R.: Domain adaptation for object recognition: an unsupervised approach. In: IEEE International Conference on Computer Vision (ICCV), pp. 999–1006 (2011)
11. Hu, L., Kan, M., Shan, S., Chen, X.: Duplex generative adversarial network for unsupervised domain adaptation. In: Proceedings of the IEEE Conference on Computer Vision and Pattern Recognition, pp. 1498–1507 (2018)
12. Kan, M., Shan, S., Chen, X.: Bi-shifting auto-encoder for unsupervised domain adaptation. In: The IEEE International Conference on Computer Vision (ICCV), pp. 3846–3854 (2015)
13. Krizhevsky, A., Sutskever, I., Hinton, G.E.: Imagenet classification with deep convolutional neural networks. In: Advances in Neural Information Processing Systems (NIPS), pp. 1097–1105 (2012)
14. LeCun, Y., Bottou, L., Bengio, Y., Haffner, P.: Gradient-based learning applied to document recognition. Proc. IEEE **86**, 2278–2324 (1998)
15. Liu, M., Tuzel, O.: Coupled generative adversarial networks. In: Advances in Neural Information Processing Systems (NIPS), pp. 469–477 (2016)
16. Liu, M.Y., Breuel, T., Kautz, J.: Unsupervised image-to-image translation networks. In: Advances in Neural Information Processing Systems (NIPS), pp. 700–708 (2017)
17. Long, M., Cao, Y., Wang, J., Jordan, M.I.: Learning transferable features with deep adaptation networks. In: International Conference on Machine learning (ICML), pp. 97–105 (2015)
18. Long, M., Zhu, H., Wang, J., Jordan, M.I.: Unsupervised domain adaptation with residual transfer networks. In: Advances in Neural Information Processing Systems (NIPS), pp. 136–144 (2016)
19. Luan, T., Yin, X., Liu, X.: Disentangled representation learning gan for pose-invariant face recognition. In: IEEE Conference on Computer vision and Pattern Recognition (CVPR), pp. 1415–1424 (2017)
20. Netzer, Y., Wang, T., Coates, A., Bissacco, A., Wu, B., Ng, A.Y.: Reading digits in natural images with unsupervised feature learning. In: Advances in Neural Information Processing Systems Workshop (NIPSW) (2011)
21. Pablo, A., Michael, M., Charless, F., Jitendra, M.: Contour detection and hierarchical image segmentation. IEEE Trans. Pattern Anal. Mach Intell. **33**(5), 898–916 (2011)
22. Pan, S.J., Tsang, I.W., Kwok, J.T., Yang, Q.: Domain adaptation via transfer component analysis. IEEE Trans. Neur. Netw. **22**(2), 199–210 (2011)
23. Perarnau, G., Van De Weijer, J., Raducanu, B., Álvarez, J.M.: Invertible conditional GANs for image editing (2016). Preprint. arXiv:1611.06355
24. Saenko, K., Kulis, B., Fritz, M., Darrell, T.: Adapting visual category models to new domains. In: European Conference on Computer Vision (ECCV), pp. 213–226 (2010)
25. Saito, K., Ushiku, Y., Harada, T.: Asymmetric tri-training for unsupervised domain adaptation. In: International Conference on Machine learning (ICML), pp. 2988–2997 (2017)
26. Sener, O., Song, H.O., Saxena, A., Savarese, S.: Learning transferrable representations for unsupervised domain adaptation. In: Advances in Neural Information Processing Systems (NIPS), pp. 2110–2118 (2016)
27. Shao, M., Castillo, C., Gu, Z., Fu, Y.: Low-rank transfer subspace learning. In: IEEE International Conference on Data Mining (ICDM), pp. 1104–1109 (2012)
28. Shao, M., Kit, D., Fu, Y.: Generalized transfer subspace learning through low-rank constraint. Int. J. Comput. Vis. **109**(1–2) (2014). https://doi.org/10.1007/s11263-014-0696-6
29. Taigman, Y., Polyak, A., Wolf, L.: Unsupervised cross-domain image generation. In: International Conference of Learning Representation (ICLR), pp. 3846–3854 (2017)
30. Tzeng, E., Hoffman, J., Saenko, K., Darrell, T.: Adversarial discriminative domain adaptation. In: IEEE Conference on Computer vision and Pattern Recognition (CVPR), pp. 7167–7176 (2017)
31. Zhu, J.Y., Park, T., Isola, P., Efros, A.A.: Unpaired image-to-image translation using cycle-consistent adversarial networks. In: IEEE Conference on Computer vision and Pattern Recognition (CVPR), pp. 2223–2232 (2017)

Chapter 7
Domain Adaptation via Image to Image Translation

Zak Murez, Soheil Kolouri, David Kriegman, Ravi Ramamoorthi, and Kyungnam Kim

7.1 Introduction

Deep convolutional neural networks (CNNs) [15, 18, 25] trained on large numbers of labeled images (tens of thousands to millions) provide state-of-the-art image representations that can be used for a wide variety of tasks including recognition, detection, and segmentation. Obtaining abundant data with high quality annotations, however, remains to be a cumbersome and expensive process in various applications. Recently, there has been a surge of interest in learning with less labels. In other words, methods that extend the unprecedented performance of deep learning methods to settings in which labeled data is scarce or not available. Few-shot learning [47], zero-shot learning [6, 48], meta-learning [38] and various unsupervised learning techniques including self-supervised learning [37] are among the topics that fall under the learning with less labels setting. In this chapter, we are interested in domain adaptation in the learning with less (in our case with zero) labels, where the learned knowledge from a source domain with abundant labeled

The authors "Zak Murez and Soheil Kolouri" contributed equally to this work.

Z. Murez · S. Kolouri (✉) · K. Kim
HRL Laboratories, LLC, Malibu, CA, USA
e-mail: skolouri@hrl.com

R. Ramamoorthi · D. Kriegman
Department of Computer Science & Engineering, University of California, San Diego (UCSD),
La Jolla, CA, USA

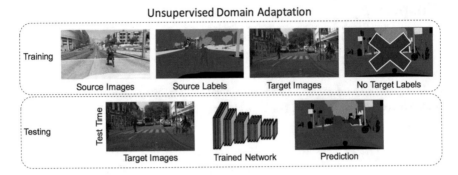

Fig. 7.1 The general unsupervised domain adaptation (UDA) setting. During training we have access to images from source and target images and source labels. However, we do not have any labels for the target dataset. The UDA algorithm then needs to learn a network that works well on both source and target domains

data to a target domain where data is unlabeled or sparsely labeled. We specifically are interested in the setting for which the target domain is completely unlabeled. This setting is known as the unsupervised domain adaptation (UDA) and the idea is depicted in Fig. 7.1.

The major challenge for knowledge transfer is a phenomenon known as domain shift [13], which refers to the different distribution of data in the target domain compared to the source domain. This problem is specifically more challenging in the UDA setting when no labels are available for the target domain, and the source model is required to generalize well on the unlabeled target data. The high-level ideas in domain adaptation are twofolds: (1) learn a mapping from the target domain to the source domain, g denoted as the adaptation map, and compose the learned model for the source domain, f, with the adaptation map, $f \circ g$, to obtain a model for the target domain, and (2) learn a shared and often low-dimensional latent space for the source and target domains, in which regression over the source domain could be performed with simple models, and the source and target representations in the latent space are indistinguishable. The latter idea is sometime referred to as a domain agnostic latent space. Generally, such mappings are not unique and there exist many mappings that align the source and target distributions (whether directly, or in a latent space). Therefore various constraints are needed to narrow down the space of feasible mappings. Recent domain adaptation techniques parameterize and learn these mappings via deep neural networks [31, 33, 44, 46, 54, 55]. In this paper, we propose a unifying, generic, and systematic framework for unsupervised domain adaptation, which is broadly applicable to many image understanding and sensing tasks where training labels are not available in the target domain. We further demonstrate that many existing methods for domain adaptation arise as special cases of our framework.

As a practical application of the domain adaptation and specifically UDA, consider the emerging application of autonomous driving where a semantic seg-

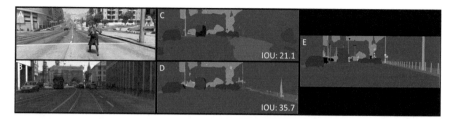

Fig. 7.2 (**a**) Sample image from the synthetic GTA5 dataset. (**b**) Input image from the real Cityscapes dataset. (**c**) Segmentation result trained on GTA5 dataset without any domain adaptation. (**d**) Ours. (**e**) Ground truth. We can see that our adaptation fixes large areas of simple mistakes on the road and sidewalk and building on the right. We also partially detect the thin pole on the right. The mean Intersection Over Union (IOU) values are reported

mentation network is required to be trained to detect traffic related objects, for instance, roads, cars, pedestrians, etc. Training such segmentation networks requires semantic, instance-wise, dense pixel annotations for each scene, which is excruciatingly expensive and time consuming to acquire. Moreover, crowd-sourced annotations often lack high quality of annotations, therefore, a large number of data-annotation-as-a-service companies have emerged in recent years to curate such annotations and provide higher quality annotations, which adds an additional cost to training reliable deep networks. To avoid human annotations, a large body of work focuses on designing photo-realistic simulated scenarios in which the ground truth annotations are readily available. Synthia [40], Virtual KITTI [8], and GTA5 [39] datasets are examples of such simulations, which include a large number of synthetically generated driving scenes together with ground truth pixel-level semantic annotations. Training a CNN based on such synthetic data and applying it to real-world images (i.e. from a dashboard mounted camera), such as the Cityscapes dataset [3], will give very poor performance due to the large differences in image characteristics which gives rise to the domain shift problem. Figure 7.2 demonstrates this scenario where a network is trained on the GTA5 dataset [39], which is a synthetic dataset, for semantic segmentation and is tested on the Cityscapes dataset [3]. It can be seen that with no adaptation the network struggles with segmentation (Fig. 7.2c), while our proposed framework ameliorates the domain shift problem and provides a more accurate semantic segmentation.

While there are significant differences between the recently developed domain adaptation methods, a common and unifying theme among these methods can be observed. We identify three main attributes needed to achieve successful unsupervised domain adaptation: (1) domain agnostic feature extraction, (2) domain specific reconstruction, and (3) cycle consistency to enable learning from unpaired source and target domains. The first requires that the distributions of features extracted from both domains are identical or as close as possible. The closeness of the source and target distributions is often measured by an adversarial network [16] or probabilistic difference measures like the Wasserstein distance [5] or sliced-Wasserstein distance [7, 24, 28]. The idea of a domain-agnostic latent space was utilized in many prior

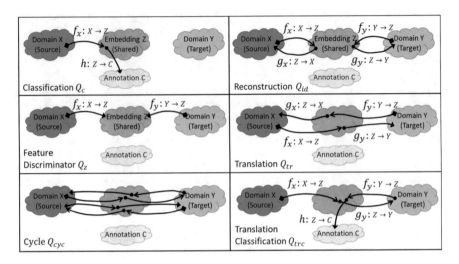

Fig. 7.3 The detailed system architecture of our I2I (image to image) Adapt framework. The pathways to the loss modules denote the inputs to these modules, which are used for training. Best viewed in color

methods [9, 10, 16], but alone does not give a strong enough constraint for domain adaptation knowledge transfer, as there exist many mappings that could match the source and target distributions in the shared space. The second is requiring that the features are able to be decoded back to the source and target domains. This idea was used in Ghifary et al. [11] for unsupervised domain adaptation and it enforces an the highly nonlinear mappings to the latent space (i.e., the deep encoders) to be pseudo-invertible, which prevents the network from collapsing areas in the input space (source or target) into single points in the latent space. Finally, the cycle consistency is needed for unpaired source and target domains to ensure that the mappings are learned correctly and they are well-behaved, in the sense that they do not collapse the distributions into single modes [59]. Figure 7.3 provides a high-level overview of our framework.

The interplay between the 'domain agnostic feature extraction', 'domain specific reconstruction with cycle consistency', and 'label prediction from agnostic features' enables our framework to simultaneously learn from the source domain and adapt to the target domain. By combining all these different components into a single unified framework we build a systematic framework for domain knowledge transfer that provides an elegant theoretical explanation as well as improved experimental results. We demonstrate the superior performance of our proposed framework for segmentation adaptation from synthetic images to real world images (see Fig. 7.2 as an example), as well as for classifier adaptation on three digit datasets. Furthermore, we show that many of the state-of-the-art methods can be viewed as special cases of our proposed framework. This chapter is partially based on earlier work from [35].

7.2   Related Work

There has been a plethora of recent work in the field of visual *domain adaptation* addressing the domain shift problem [13]. Not surprisingly, the majority of recent work utilize deep convolutional architectures to overcome the 'domain gap,' i.e. the difference between source and target domain distributions. A common approach is to map the source and target domains into a shared space (i.e., a kernel space) where the domains are aligned [16, 26, 44, 53–55]. These methods widely differ on the type of nonlinear mappings to the kernel/latent space as well as the choices of loss functions used for training them. The main theme behind the majority of these work is to embed the source and target data in a shared latent space, in which the discrepancy between source and target distributions is minimized. Various work have used different measures to minimize this discrepancy. For instance, some have used Maximum Mean Discrepancy (MMD) between the distributions of the source and target domains in the shared space [31]. The Wasserstein distance and its variations [4, 5, 7, 28, 41] have also been used for measuring source and target discrepancy, while other methods have used correlation maximization to align the second-order statistics of the domains.

Alternatively, in order to match the distribution of source and target domains one could maximize the confusion rate of an adversarial network, that is required to distinguish the source and target domains in the shared space [9–11, 16, 54]. This leads to the field of adversarial learning, which was introduced in the seminal work of Goodfellow et al. [12] and has become quite popular in the recent years. We note that, adversarial learning could be challenging to train in the sense that achieving a Nash equilibrium between the source and target encoders and the discriminator (i.e., the adversary) is not guaranteed. In addition, naive adversarial learning approaches may suffer from known problems such as the mode collapse issue [49]. There has been a large body of work on improving adversarial learning to provide more reliable training [1, 14] and address the mode collapse problem [49]. Other approaches include the work by Sener et al. [45], where the domain transfer is formulated in a transductive setting, and the Residual Transfer Learning (RTL) approach [32] was used where the authors assume that the source and target classifiers only differ by a residual function and learn these residual functions.

Another important strategy for unsupervised domain adaptation is based on self-training [52, 60]. While there are various flavors of self-training, the general idea is to enforce label consistencies between predicted target labels/annotations, i.e., pseudo-labels/annotations of the target domain, and the ground truth source labels. To provide a more tangible example, consider the semantic segmentation problem in the autonomous driving applications. A model trained on the source domain and tested on the target domain would provide poor semantic segmentation (compare Fig. 7.2c and e). It is quite simple for a human to distinguish the ground-truth labels from the predicted labels, due to the irregularities in the predicted labels. The idea is then to train an adversarial network to distinguish the predicted annotations from the ground truth (at a pixel level), and maximize the confusion of such network to perform domain adaptation. The same idea could be used via minimizing probabilistic distances between the predicted labels of the source and target domains [7, 28].

The method presented in the chapter is primarily motivated by the work of Hoffman et al. [16], Isola et al. [20], Zhu et al. [59], and Ghifary et al. [11]. Hoffman et al. [16] utilized fully convolutional networks with domain adversarial training to obtain domain agnostic features (i.e. shared space) for the source and target domains, while constraining the shared space to be discriminative for the source domain. Hence, by learning the mappings from source and target domains to the shared space (i.e. f_x and f_y in Fig. 7.3), and learning the mapping from the shared space to annotations (i.e. h in Fig. 7.3), their approach effectively enables the learned classifier to be applicable to both domains. The Deep Reconstruction Classification Network (DRCN) of Ghifary et al. [11], utilizes a similar approach but with a constraint that the embedding must be decodable, and learns a mapping from the embedding space to the target domain (i.e. g_y in Fig. 7.3). The image-to-image translation work by Isola et al. [20] maps the source domain to the target domain by an adversarial learning of f_x and g_y and composing them $g_y \circ f_x : X \rightarrow Y$. In their framework the target and source images were assumed to be paired, in the sense that for each source image there exists a known corresponding target image. This assumption was lifted in the follow-up work of Zhu et al. [59] and Royer et al. [42], where cycle consistency was used to learn the mappings based on unpaired source and target images. While the approaches of Isola et al. [20] and Zhu et al. [59] do not address the domain adaptation problem, yet they provide a baseline for learning high quality mappings from a visual domain into another.

The patterns that collectively emerge from the mentioned papers [11, 16, 20, 53, 59], are: (a) the shared space must be a discriminative embedding for the source domain, (b) the embedding must be domain agnostic, hence maximizing the similarity between the distributions of embedded source and target images, (c) the information preserved in the embedding must be sufficient for reconstructing domain specific images, (d) adversarial learning as opposed to the classic losses can significantly enhance the quality of learned mappings, (e) cycle-consistency is required to reduce the space of possible mappings and ensure their quality, when learning the mappings from unpaired images in the source and target domains. Our proposed method for unsupervised domain adaptation unifies the above-mentioned pieces into a generic framework that simultaneously solves the domain adaptation and image-to-image translation problems.

There have been other recent efforts toward a unifying and general framework for deep domain adaptation. The Adversarial Discriminative Domain Adaptation (ADDA) work by Tzeng et al. [55] and the Generate to Adapt work by Sankaranarayanan et al. [44] are instances of such frameworks. Tzeng et al. [55] identify three design choices for a deep domain adaptation system, namely (a) whether to use a generative or discriminative base, whether to share mapping parameters between f_x and f_y, and the choice of adversarial training. They observed that modeling image distributions might not be strictly necessary if the embedding is domain agnostic (i.e. domain invariant). Sankaranarayanan et al. [44] propose a different framework in which they strongly enforce the domain alignment in the shared space by learning g_y to be a generative model that maps the encoded domain images, $f_x(x_i)$, to the source domain. Our framework also generalizes these frameworks. Finally, we note that very similar ideas to ours have been proposed concurrently by Hoffman et al. [17] and Liu et al. [30].

7.3 Method

Here we denote training images with $x_i \in X$, where $X \subset \mathbb{R}^{d_x}$, and their corresponding annotations/labels $c_i \in C$, where $C \subset \mathbb{R}^K$ denotes the label/annotation space, from the source domain (i.e. domain X). Note that c_i may be image level such as in classification for which c_i denote the class assignment to one of the K classes, or pixel level in the case of semantic segmentation, where $c_i[m, n]$ identify the class assignment of a pixel $[m, n]$ in image x_i to K classes (e.g., pedestrian, road, sidewalk, etc.). Also consider training images $y_j \in Y$ in the target domain (i.e. domain $Y \subset \mathbb{R}^{d_y}$), where we do not have corresponding annotations for these images.

Our goal is then to learn a classifier that maps the target images, y_js, to labels $c_j \in C$. We note that the framework is readily extensible to a semi-supervised learning or few-shot learning scenario where we have annotations for a few images in the target domain. Given that the target domain lacks labels, the general approach is to learn a classifier on the source domain and adapt it in a way that its domain distribution matches that of the target domain. Note that, we assume no correspondences between images in the source and target domains, i.e., x_i and y_js. The lack of correspondences assumption makes the problem significantly more challenging and has drawn significant attention from the computer vision community in recent years [17].

The overarching idea here is to find a joint latent space, $Z \subset \mathbb{R}^{d_z}$ where more often than not $d_z \ll d_x, d_y$, for the source and target domains, X and Y, where the representations are domain agnostic. To clarify this point, consider the scenario in which X is the domain of driving scenes/images on a sunny day and Y is the domain of driving scenes on a rainy day. While 'sunny' and 'rainy' are characteristics of the source and target domains, they are truly nuisance variations with respect to the annotation/classification task (e.g. semantic segmentation of the road), as they should not affect the annotations. Treating such characteristics as structured noise, we would like to find a latent space, Z, that is invariant to such variations. In other words, domain Z should not contain domain specific characteristics, hence it should be domain agnostic. In what follows we describe the process that leads to finding such a domain agnostic latent space.

Let the mappings from source and target domains to the latent space be defined as $f_x : X \rightarrow Z$ and $f_y : Y \rightarrow Z$, respectively (see Fig. 7.3). In our framework these mappings are parameterized by deep convolutional neural networks (CNNs). Here it should be noted that the mappings f_x and f_y could have shared parameters. For instance, when X is equal to Y (i.e., $d_x = d_y$) and the input images share similar low level visual features (e.g., the source and target images are both RGB images of road scenes, however, at different weather conditions), then f_x and f_y could share all their parameters, or $f_x = f_y$. In more general settings, for instance for RGB to Synthetic Aperture Radar (SAR) domain adaptation as in [41], the mappings f_x and f_y could only share parameters at deeper layers of the CNNs, or not share parameters at all. Not sharing any parameters between f_x and f_y, however, could lead to less robust domain adaptation, as it allows the highly nonlinear mappings f_x and f_y to warp the source and target spaces and provide arbitrary matchings between X and Y in the latent space Z. Note that the members of the latent space $z \in Z$ are high dimensional vectors in the case of image level tasks, or feature maps in the case of pixel level tasks. Also, let $h : Z \rightarrow C$ be the classifier that maps the latent space to labels/annotations (i.e. the classifier module in Fig. 7.3). Given that the annotations for the source class X are known, one can define a supervised loss function to enforce $h(f_x(x_i)) = c_i$:

$$Q_c = \sum_i l_c \left(h(f_x(x_i)), c_i\right) \tag{7.1}$$

where l_c is an appropriate loss (e.g. cross entropy for classification and segmentation). Minimizing the above loss function leads to the standard approach of supervised learning, which does not concern domain adaptation. While this approach would lead to a method that performs well on the images in the source domain, $x_i \in X$, it will more often than not perform poorly on images from the target domain $y_j \in Y$. The reason is that, domain Z is heavily biased to the distribution of the structured noise ('sunny') in domain X and the structured noise in domain Y ('rainy') confuses the classifier $h(\cdot)$. To avoid such confusion we require the latent space, Z, to be domain agnostic, so it is not sensitive to the domain specific structured noise. To achieve such a latent space we systematically introduce a variety of auxiliary networks and losses to help regularize the latent space and consequently achieve a robust $h(\cdot)$. The auxiliary networks and loss pathways are depicted in Fig. 7.3. In what follows we describe the individual components of the regularization losses (Fig. 7.4).

1. First of all, Z is required to preserve the core information of the target and source images and only discard the structured noise. To impose this constraint on the latent space, we first define decoders $g_x : Z \rightarrow X$ and $g_y : Z \rightarrow Y$ that take the features in the latent space to the source and target domains, respectively. In other words, g_x and g_y are the pseudo-inverses of f_x and f_y, respectively. We assume that if Z retains the crucial/core information of the domains and only discards the structured noise, then the decoders should be able to add the structured noise back and reconstruct each image from their representation in the latent feature space, Z. In other words, we require $g_x(f_x(\cdot))$ and $g_y(f_y(\cdot))$ to be close to identity functions/maps ($f_x \circ g_x \approx id$). This constraint leads to the following loss function:

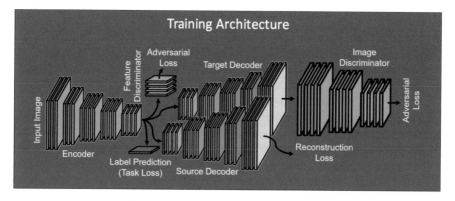

Fig. 7.4 Our proposed training architecture

$$Q_{id} = \sum_i l_{id}\left(g_x(f_x(x_i)), x_i\right) +$$

$$\sum_j l_{id}\left(g_y(f_y(y_j)), y_j\right) \tag{7.2}$$

where $l_{id}(\cdot, \cdot)$ is a pixel-wise image loss such as the L_1 norm.

2. We would like the latent space Z to be domain agnostic. This means that the feature representations of the source and target domain should not contain domain specific information. To achieve this, we use an adversarial setting in which a discriminator $d_z : Z \rightarrow \{c_x, c_y\}$ tries to classify if a feature in the latent space $z \in Z$ was generated from domain X or Y, where c_x and c_y are binary domain labels (i.e. from domain X or domain Y). The loss function then can be defined as the certainty of the discriminator (i.e. domain agnosticism is equivalent to fooling the discriminator), and therefore we can formulate this as:

$$Q_z = \sum_i l_a\left(d_z(f_x(x_i)), c_x\right) +$$

$$\sum_j l_a\left(d_z(f_y(y_j)), c_y\right) \tag{7.3}$$

where $l_a(\cdot, \cdot)$ is an appropriate loss (the cross entropy loss in traditional GANs [12] and mean square error in least squares GAN [34]). The discriminator is trained to maximize this loss while the discriminator is trained to minimize it. Note that, the core idea behind using a discriminator d_z and adversarial learning here is to minimize the discrepancy between the distribution of $f_x(x_i)$ and $f_y(y_j)$ in the latent space. To that end, other approaches have been introduced that use statistical difference measures between empirical probability distributions of source and target domains in the latent space, i.e., $p_{f_x} = \frac{1}{N}\sum_{i=1}^{N}\delta(z - f_x(x_i))$

and $p_{f_y} = \frac{1}{N} \sum_{j=1}^{N} \delta(z - f_y(y_j))$ where $\delta(\cdot)$ is the Dirac function. For instance, Damodaran et al. [5] used the Wasserstein distance to achieve this task, while [7, 24, 28, 41] use the sliced-Wasserstein distance to address the computational complexity of calculating the Wasserstein distance itself. For a more comprehensive introduction to the Wasserstein distances please refer to Kolouri et al. [23] and Arjovsky et al. [1]. Other notable discrepancy measures, include the maximum-mean-discrepancy (MMD) measure [31] and the Cramer-Wold metric [50].

3. To further ensure that the mappings f_x, f_y, g_x, and g_y are consistent we define translation adversarial losses. An image from target (source) domain is first encoded to the latent space and then decoded to the source (target) domain to generate a 'fake' (translated) image. Next, we define discriminators $d_x : X \rightarrow \{c_x, c_y\}$ and $d_y : Y \rightarrow \{c_x, c_y\}$, to identify if an image is 'fake' (generated from the other domain) or 'real' (belonged to the actual domain). To formulate this translation loss function we can write:

$$Q_{tr} = \sum_i l_a \left(d_y(g_y(f_x(x_i))), c_x \right) +$$

$$\sum_j l_a \left(d_x(g_x(f_y(y_j))), c_y \right) \tag{7.4}$$

4. Given that there are no correspondences between the images in the source and target domains, we need to ensure that the semantically similar images in both domains are projected into close vicinity of one another in the latent space. To ensure this, we define the cycle consistency losses where the 'fake' images generated in the translation loss, $g_x(f_y(y_j))$ or $g_y(f_x(x_i))$, are encoded back to the latent space and then decoded back to their original space. The entire cycle should be equivalent to an identity mapping. We can formulate this loss as follows:

$$Q_{cyc} = \sum_i l_{id} \left(g_x(f_y(g_y(f_x(x_i)))), x_i \right) +$$

$$\sum_j l_{id} \left(g_y(f_x(g_x(f_y(y_j)))), y_j \right) \tag{7.5}$$

5. To further constrain the translations to maintain the same semantics, and allow the target encoder to be trained with supervision on target domain 'like' images we also define a classification loss between the source to target translations and the original source labels:

$$Q_{trc} = \sum_i l_c \left(h(f_y(g_y(f_x(x_i)))), c_i \right) \tag{7.6}$$

Table 7.1 Showing the relationship between the existing methods and our proposed method

Method	Q_c	Q_z	Q_{tr}	Q_{id_X}	Q_{id_Y}	Q_{cyc}	Q_{trc}
[16]	✓	✓					
[55]	✓	✓					
[11]	✓				✓		
[44]	✓		✓				
[59]			✓			✓	
Ours	✓	✓	✓	✓	✓	✓	✓

Finally, by combining these individual losses we define the general loss to be,

$$Q = \lambda_c Q_c + \lambda_z Q_z + \lambda_{tr} Q_{tr} + \lambda_{id} Q_{id} + \lambda_{cyc} Q_{cyc} + \lambda_{trc} Q_{trc} \qquad (7.7)$$

A variety of prior methods for domain adaptation are special cases of our framework. Table 7.1 summarizes which hyperparameters to include and which are set to zero to recover these prior methods. In short, by setting $\lambda_{id} = \lambda_{cyc} = \lambda_{tr} = 0$ we recover [16]. By first training only on the source domain and then freezing the source encoder, untying the target encoder and setting $\lambda_{id} = \lambda_{cyc} = \lambda_{tr} = 0$ we recover [55]. By setting $\lambda_{id_X} = \lambda_{cyc} = \lambda_{tr} = \lambda_z = 0$ we recover [11], where λ_{id_X} indicates the mixing coefficient only for the first term of Q_{id}. Setting $\lambda_{id} = \lambda_{cyc} = \lambda_{tr_X} = \lambda_z = 0$ we recover [44]. Finally, by setting $\lambda_{id} = \lambda_c = \lambda_z = 0$ we recover [59]. Table 7.1 summarizes these results.

7.4 Experiments

In this section we demonstrate the performance of the described method on various benchmark domain adaptation datasets, including: (1) digit recognition datasets consisting of MNIST, USPS, and SVHN datasets, (2) Office dataset, and (3) GTA5 and Cityscapes datasets. The first two adaptation tasks are classification tasks, while the last one is a segmentation task. In all our experiments, the loss function is optimized via the ADAM optimizer [22], which provides adaptive gradient steps based on lower-order moments, with learning rate 0.0002 and betas 0.5 and 0.999, in an end-to-end manner. The discriminative networks, d_x, d_y, and d_z are trained in an alternating optimization scheme alongside with the encoders and decoders.

To further constrain the features that are learned we share the weights of the encoders. We also share the weights of the first few layers of the decoders. To stabilize the discriminators we train them using the Improved Wasserstein method [14]. The loss of the feature discriminator (Q_z) is only backpropagated to the generator for target images (we want the encoder to learn to map the target images to the same distribution as the source images, not vice versa). Likewise, the translation classification loss (Q_{trc}) is only backpropagated to the second encoder and classifier (f_y and h). This prevents the translator (g_y) from cheating by hiding class information in the translated images. In what follows we present the results of our method on the above mentioned datasets.

7.4.1 MNIST, USPS, and SVHN Digits Datasets

First, we demonstrate our method on domain adaptation between three digit classification datasets, namely MNIST [27], USPS [19], and the Street View House Numbers (SVHN) [36] datasets. We followed the experimental protocol of [10, 29, 44, 54, 55] where we treated one of the digit datasets as a labeled source domain and another dataset as unlabeled target domain. We trained our framework for adaptation from MNIST→ USPS, USPS → MNIST, and SVHN → MNIST. Figure 7.5 shows examples of MNIST to SVHN input and translated images.

For a fair comparison with previous methods, our feature extractor network (encoder, f_x and f_y) is a modified version of LeNet [27]. Our modified LeNet encoder consists of 4 stride 2 convolutional layers with 4×4 filters and 64, 64, 128, 128 features respectively. Each convolution is followed by batch normalization and a ReLU nonlinearity. Batch normalization was helpful with the image to image

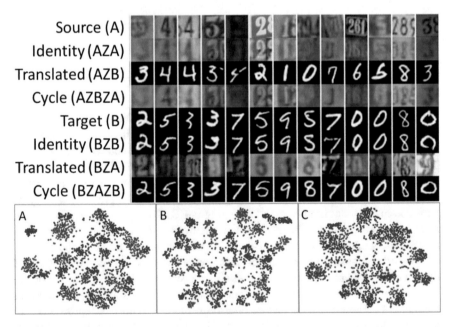

Fig. 7.5 (Top) Image to image translation examples for MNIST to SVHN. (Bottom) TSNE embedding visualization of the latent space. Red are source images, Blue are target images. (**a**) No adaptation. (**b**) Image to image adaptation without latent space discriminator. (**c**) Full adaptation

translation training. All weights are shared between the source and target encoders. Our DenseNet encoder follows [18] with the final fully connected layer removed.

Our decoders (i.e. g_x and g_y) consist of four stride 2 transposed convolutional layers with 4×4 filters and 512, 256, 128, 1 features respectively. Each convolution is followed by batch normalization and a ReLU nonlinearity except the last layer which only has a Tanh nonlinearity. The weights of the first two layers are shared between the source and target decoders. Our image discriminators consist of 4 stride 2 convolutional layers with 4×4 filters and 64, 128, 256, 1 features respectively, and our feature discriminator consists of three fully connected layers. More precisely, the feature discriminator consists of 3 linear layers with 500, 500, 1 features, each followed by a leaky ReLU nonlinearity with slope 0.2. The feature discriminator is trained with the Least Squares GAN loss. The Loss is only backpropagated to the generator for target images (we want the encoder to learn to map the target images to the same distribution as the source images, not vice versa). All images from MNIST and USPS were bilinearly upsampled to 32×32. Images from SVHN were converted to gray scale. We also included a very simple data augmentation in the form of random translations and rotations. We found out that data augmentation is a crucial step, which helps with increasing the domain adaptation performance significantly. For our hyperparameters we used: $\lambda_c = 1.0$, $\lambda_z = 0.05$, $\lambda_{id} = 0.1$, $\lambda_{tr} = 0.02$, $\lambda_{cyc} = 0.1$, $\lambda_{trc} = 0.1$.

We compare our method to nine prior works (see Table 7.2). Our method consistently out performs the prior state of the art by a significant margin. We also show ablations to analyze how much each of the loss terms contributes to the overall performance. First note that even our results without domain adaption (top row of Table 7.2 below the line) are better than many prior methods. This is purely due to the simple data augmentation. Next we see that λ_z and λ_{id_y} both improve results a lot. However the combination of them with λ_{trc} produces the best results. This is especially apparent in SVHN to MNIST, which is among the most challenging adaptation tasks. Finally, in this case, the remaining losses don't help any further.

Figure 7.5a–c show TSNE embeddings of the features extracted from the source and target domain when trained without adaptation, with image to image loss only, and our full model. It can be seen that without adaptation, the source and target images get clustered in the feature space but the distributions do not overlap which is why classification fails on the target domain. Just image to image translation is not enough to force the distributions to overlap as the networks learn to map source and target distributions to different areas of the feature space. Our full model includes a feature distribution adversarial loss, forcing the source and target distributions to overlap, while image translation makes the features richer yielding the best adaptation results.

Table 7.2 Performance of various prior methods as well as ours and ablations on digits datasets domain adaptation

Method						MNIST → USPS	USPS → MNIST	SVHN → MNIST
Source only						24.8	42.9	39.7
Gradient reversal [10]						28.9	27.0	26.1
Domain confusion [54]						20.9	33.5	31.9
CoGAN [29]						8.8	10.9	–
ADDA [55]						10.6	9.9	24.0
DTN [51]						–	–	15.6
WDAN [57]						27.4	34.6	32.6
PixelDA [2]						4.1	–	–
DRCN [11]						8.2	26.3	18.0
Gen to adapt [44]						7.5	9.2	15.3
Q_z	Q_{tr}	Q_{id_X}	Q_{id_Y}	Q_{cyc}	Q_{trc}	Ours		
						8.9	33.0	28.5
✓						2.1	2.8	19.9
			✓			2.1	28.5	23.7
			✓		✓	1.4	22.6	29.5
✓			✓			1.3	3.0	12.1
✓			✓		✓	**1.2**	**2.4**	**9.9**
✓	✓	✓	✓		✓	2.5	3.6	10.0
✓	✓	✓	✓	✓	✓	1.5	2.6	10.4

MNIST → USPS indicates MNIST is the source domain (labels available) and USPS is the target domain (no labels available). Results reported are classification error rate (lower is better). Blue is best prior method, bold is best overall. Our results are considerably better than the prior state of the art

7.4.2 Office Dataset

The Office dataset [43] consists of images from 31 classes of objects in three domains: Amazon (A), Webcam (W) and DSLR (D) with 2817, 795 and 498 images respectively. Our method performs the best in four out of six of the tasks (see Table 7.3). The two tasks that ours is not best at consist of bridging a large domain shift with very little training data in the source domain (795 and 498 respectively). Here the ablations show that the translation loss (Q_{tr}) helps.

For our encoder we use a ResNet34 pretrained on ImageNet. The encoder is trained with a smaller learning rate (2×10^{-5}), to keep the weights closer to their good initialization. Images are down sampled to 256×256 and then a random crop of size 224×244 is extracted. The final classification layer is applied after global average pooling. Our decoders consist of 5 stride 2 transposed convolutional layers. The image discriminators consist of 4 stride 2 convolutional layers. The feature discriminator consists of 3 1×1 convolutions followed by global average pooling.

Our hyperparameters were: $\lambda_c = 1.0$, $\lambda_z = 0.1$, $\lambda_{tr} = 0.005$, $\lambda_{id} = 0.2$, $\lambda_{cyc} = 0.0$, $\lambda_{trc} = 0.1$.

Table 7.3 Accuracy (larger is better) of various methods on the Office dataset consisting of three domains: Amazon (A), Webcam (W) and DSLR (D)

Method	A → W	W → A	A → D	D → A	W → D	D → W
Domain confusion [53]	61.8	52.2	64.4	21.1	98.5	95.0
Transferable features [31]	68.5	53.1	67.0	54.0	99.0	96.0
Gradient reversal [10]	72.6	52.7	67.1	54.5	99.2	96.4
DHN [56]	68.3	53.0	66.5	55.5	98.8	96.1
WDAN [57]	66.8	52.7	64.5	53.8	98.7	95.9
DRCN [11]	68.7	**54.9**	66.8	**56.0**	99.0	96.4

Q_z	Q_{tr}	Q_{id_X}	Q_{id_Y}	Q_{trc}	Ours	A → W	W → A	A → D	D → A	W → D	D → W
						59.1	46.4	61.0	45.3	98.0	92.8
✓						70.8	49.0	67.1	43.4	98.2	90.8
		✓				61.1	49.6	67.3	49.8	99.0	94.7
✓			✓	✓		71.2	49.1	70.9	45.5	97.8	94.3
✓	✓	✓	✓	✓		**75.3**	52.1	**71.1**	50.1	**99.6**	**96.5**

A → W indicates Amazon is the source domain (labels available) and Webcam is the target domain (no labels available). Bold is best. Our method performs best on 4 out of 6 of the tasks

7.4.3 GTA5 to Cityscapes

We also demonstrate our method for domain adaptation between the synthetic (photorealistic) driving dataset GTA5 [39] and the real dataset Cityscapes [3]. The GTA5 dataset consists of 24,966 densely labeled RGB images of size 1914×1052, containing 19 classes that are compatible with the Cityscapes dataset (see Table 7.4). The Cityscapes dataset contains 5000 densely labeled RGB images of size 2040×1016 from 27 different cities. Here the task is pixel level semantic segmentation. Following the experiment in [16], we use the GTA5 images as the labeled source dataset and the Cityscapes images as the unlabeled target domain.

We point out that the convolutional networks in our model are interchangeable. We include results using a dilated ResNet34 encoder for fair comparison with previous work, but we found from our experiments that the best performance was achieved by using our new Dilated Densely-Connected Networks (i.e. Dilated DenseNets) for the encoders which are derived by replacing strided convolutions with dilated convolutions [58] in the DenseNet architecture [18]. DenseNets have previously been used for image segmentation [21] but their encoder/decoder structure is more cumbersome than what we proposed. We use a series of transposed convolutional layers for the decoders.

Our decoders consist of a stride 1 convolutional layer followed by 3 stride 2 transposed convolutional layers. The image discriminators consist of 4 stride 2 convolutional layers. We did not include the cycle consistency constraint due to memory issues. Due to computational and memory constraints, we down sample all images by a factor of two prior to feeding them into the networks. Output segmentations are bilinearly up sampled to the original resolution. We train our network on 256×256 patches of the down sampled images, but test on the full

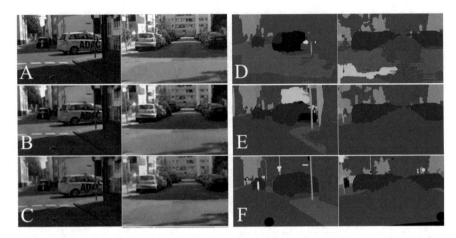

Fig. 7.6 (**a**) Input image from real Cityscapes dataset. (**b**) Identity mapped image. (**c**) Translated image. (**d**) Segmentation without domain adaptation. (**e**) Our Segmentation. (**f**) Ground truth. Although our image translations might not be as visually pleasing as those in [59] (our architecture is not optimized for translation), they succeed in their goal of domain adaptation

(downsampled) images convolutionally. Our hyperparameters were: $\lambda_c = 1.0$, $\lambda_z = 0.01$, $\lambda_{tr} = 0.04$, $\lambda_{id} = 0.2$, $\lambda_{cyc} = 0.0$, $\lambda_{trc} = 0.1$.

Our encoder architecture (dilated ResNet/DenseNet) is optimized for segmentation and thus it is not surprising that our translations (see Fig. 7.6) are not quite as good as those reported in [59]. Qualitatively, it can be seen from Fig. 7.6 that our segmentations are much cleaner compared to no adaptation. Quantitatively (see Table 7.4), our method outperforms the previous method [16] on all categories except 3, and is 5% better overall. Furthermore, we show that using Dilated DenseNets in our framework, increases the SOA by 8.6%.

7.5 Final Remarks

In this book chapter, we proposed a general framework for unsupervised domain adaptation (UDA) which encompasses many recent works as special cases. We provided a comprehensive review of the existing work in the literature and brought the recent progress in the field of UDA under a unifying theoretical umbrella. Our proposed method simultaneously achieves image to image translation, source discrimination, and domain adaptation and provide a roadmap for follow-up work on UDA. In addition, the proposed method is readily extendable to the semi-supervised and few-shot learning frameworks. Our implementation outperforms state of the art on adaptation for digit classification and semantic segmentation of driving scenes. When combined with the DenseNet architecture our method significantly outperforms the current state of the art.

Table 7.4 Performance (Intersection over Union) of various methods on driving datasets domain adaptation

| Category | Methods | | | | |
	Source	FCNs in the wild	Ours	Source—DenseNet	Ours—DenseNet
Road	31.9	67.4	85.3	67.3	**85.8**
Sidewalk	18.9	29.2	**38.0**	23.1	37.5
Building	47.7	64.9	71.3	69.4	**80.2**
Wall	7.4	15.6	18.6	13.9	**23.3**
Fence	3.1	8.4	16.0	14.4	**16.1**
Pole	16.0	12.4	18.7	21.6	**23.0**
Traffic light	10.4	9.8	12.0	**19.2**	14.5
Traffic sign	1.0	2.7	4.5	**12.4**	9.8
Vegetation	76.5	74.1	72.0	78.7	**79.2**
Terrain	13.0	12.8	43.4	24.5	**36.5**
Sky	58.9	66.8	63.7	74.8	**76.4**
Person	36.0	38.1	43.1	49.3	**53.4**
Rider	1.0	2.3	3.3	3.7	**7.4**
Car	67.1	63.0	76.7	54.1	**82.8**
Truck	9.5	9.4	14.4	8.7	**19.1**
Bus	3.7	5.1	12.8	5.3	**15.7**
Train	0.0	0.0	0.3	2.6	**2.8**
Motorcycle	0.0	3.5	9.8	6.2	**13.4**
Bicycle	0.0	0.0	0.6	**1.9**	1.7
mIoU	21.1	27.1	31.8	29.0	**35.7**

Above the line uses the standard dilated ResNet as the encoder. Our method performs the best overall and on all sub categories except two. Switching to a DenseNet encoder beats the previous method even without domain adaptation. DenseNet plus our method significantly out performs the previous method. Blue is best with ResNet, Bold is best overall

The three take-away concepts from this chapter are the importance of: (1) domain agnostic feature extraction, (2) domain specific reconstruction, and (3) cycle consistency to enable learning from unpaired source and target domains, for unsupervised domain adaptation.

References

1. Arjovsky, M., Chintala, S., Bottou, L.: Wasserstein generative adversarial networks. In: International Conference on Machine Learning, pp. 214–223 (2017)
2. Bousmalis, K., Silberman, N., Dohan, D., Erhan, D., Krishnan, D.: Unsupervised pixel-level domain adaptation with generative adversarial networks. In: Proceedings of the IEEE Conference on Computer Vision and Pattern Recognition, pp. 3722–3731 (2017)
3. Cordts, M., Omran, M., Ramos, S., Rehfeld, T., Enzweiler, M., Benenson, R., Franke, U., Roth, S., Schiele, B.: The cityscapes dataset for semantic urban scene understanding. In: Proc. of the IEEE Conference on Computer Vision and Pattern Recognition (CVPR) (2016)

4. Courty, N., Flamary, R., Tuia, D., Rakotomamonjy, A.: Optimal transport for domain adaptation. IEEE Trans. Pattern Anal. Mach. Intell. **39**(9), 1853–1865 (2016)
5. Damodaran, B.B., Kellenberger, B., Flamary, R., Tuia, D., Courty, N.: Deepjdot: deep joint distribution optimal transport for unsupervised domain adaptation. In: European Conference on Computer Vision, pp. 467–483. Springer, Berlin (2018)
6. Deutsch, S., Kolouri, S., Kim, K., Owechko, Y., Soatto, S.: Zero shot learning via multi-scale manifold regularization. In: Proceedings of the IEEE Conference on Computer Vision and Pattern Recognition, pp. 7112–7119 (2017)
7. Gabourie, A., Rostami, M., Pope, P., Kolouri, S., Kim, K.: Learning a domain-invariant embedding for unsupervised domain adaptation using class-conditioned distribution alignment (2019). Preprint, arXiv:1907.02271
8. Gaidon, A., Wang, Q., Cabon, Y., Vig, E.: Virtual worlds as proxy for multi-object tracking analysis. In: Proceedings of the IEEE Conference on Computer Vision and Pattern Recognition, pp. 4340–4349 (2016)
9. Ganin, Y., Lempitsky, V.: Unsupervised domain adaptation by backpropagation. In: International Conference on Machine Learning, pp. 1180–1189 (2015)
10. Ganin, Y., Ustinova, E., Ajakan, H., Germain, P., Larochelle, H., Laviolette, F., Marchand, M., Lempitsky, V.: Domain-adversarial training of neural networks. J. Mach. Learn. Res. **17**(59), 1–35 (2016)
11. Ghifary, M., Kleijn, W.B., Zhang, M., Balduzzi, D., Li, W.: Deep reconstruction-classification networks for unsupervised domain adaptation. In: European Conference on Computer Vision, pp. 597–613. Springer, Berlin (2016)
12. Goodfellow, I., Pouget-Abadie, J., Mirza, M., Xu, B., Warde-Farley, D., Ozair, S., Courville, A., Bengio, Y.: Generative adversarial nets. In: Advances in Neural Information Processing Systems, pp. 2672–2680 (2014)
13. Gretton, A., Smola, A.J., Huang, J., Schmittfull, M., Borgwardt, K.M., Schölkopf, B.: Covariate shift by kernel mean matching (2009)
14. Gulrajani, I., Ahmed, F., Arjovsky, M., Dumoulin, V., Courville, A.: Improved training of Wasserstein GANs (2017). Preprint, arXiv:1704.00028
15. He, K., Zhang, X., Ren, S., Sun, J.: Deep residual learning for image recognition. In: Proceedings of the IEEE Conference on Computer Vision and Pattern Recognition, pp. 770–778 (2016)
16. Hoffman, J., Wang, D., Yu, F., Darrell, T.: FCNs in the wild: pixel-level adversarial and constraint-based adaptation (2016). Preprint, arXiv:1612.02649
17. Hoffman, J., Tzeng, E., Park, T., Zhu, J.Y., Isola, P., Saenko, K., Efros, A.A., Darrell, T.: Cycada: cycle-consistent adversarial domain adaptation (2017). Preprint, arXiv:1711.03213
18. Huang, G., Liu, Z., Weinberger, K.Q., van der Maaten, L.: Densely connected convolutional networks (2016). Preprint, arXiv:1608.06993
19. Hull, J.J.: A database for handwritten text recognition research. IEEE Trans. Pattern Anal. Mach. Intell. **16**(5), 550–554 (1994)
20. Isola, P., Zhu, J.Y., Zhou, T., Efros, A.A.: Image-to-image translation with conditional adversarial networks (2016). Preprint, arXiv:1611.07004
21. Jégou, S., Drozdzal, M., Vazquez, D., Romero, A., Bengio, Y.: The one hundred layers tiramisu: Fully convolutional densenets for semantic segmentation. In: IEEE Conference on Computer Vision and Pattern Recognition Workshops (CVPRW), 2017, pp. 1175–1183. IEEE, Piscataway (2017)
22. Kingma, D.P., Ba, J.: Adam: a method for stochastic optimization (2014). Preprint, arXiv:1412.6980
23. Kolouri, S., Park, S.R., Thorpe, M., Slepcev, D., Rohde, G.K.: Optimal mass transport: Signal processing and machine-learning applications. IEEE Signal Process. Mag. **34**(4), 43–59 (2017)
24. Kolouri, S., Pope, P.E., Martin, C.E., Rohde, G.K.: Sliced Wasserstein auto-encoders. In: International Conference on Learning Representations (2019). https://openreview.net/forum?id=H1xaJn05FQ

25. Krizhevsky, A., Sutskever, I., Hinton, G.E.: Imagenet classification with deep convolutional neural networks. In: Advances in Neural Information Processing Systems, pp. 1097–1105 (2012)
26. Kulis, B., Saenko, K., Darrell, T.: What you saw is not what you get: domain adaptation using asymmetric kernel transforms. In: CVPR 2011, pp. 1785–1792. IEEE, Piscataway (2011)
27. LeCun, Y., Bottou, L., Bengio, Y., Haffner, P.: Gradient-based learning applied to document recognition. Proc. IEEE **86**(11), 2278–2324 (1998)
28. Lee, C.Y., Batra, T., Baig, M.H., Ulbricht, D.: Sliced Wasserstein discrepancy for unsupervised domain adaptation (2019). Preprint, arXiv:1903.04064
29. Liu, M.Y., Tuzel, O.: Coupled generative adversarial networks. In: Advances in Neural Information Processing Systems, pp. 469–477 (2016)
30. Liu, M.Y., Breuel, T., Kautz, J.: Unsupervised image-to-image translation networks. In: Advances in Neural Information Processing Systems, pp. 700–708 (2017)
31. Long, M., Cao, Y., Wang, J., Jordan, M.: Learning transferable features with deep adaptation networks. In: International Conference on Machine Learning, pp. 97–105 (2015)
32. Long, M., Zhu, H., Wang, J., Jordan, M.I.: Unsupervised domain adaptation with residual transfer networks. In: Advances in Neural Information Processing Systems, pp. 136–144 (2016)
33. Luo, Z., Zou, Y., Hoffman, J., Fei-Fei, L.: Label efficient learning of transferable representations across domains and tasks. In: Conference on Neural Information Processing Systems (NIPS) (2017)
34. Mao, X., Li, Q., Xie, H., Lau, R.Y., Wang, Z.: Multi-class generative adversarial networks with the l2 loss function (2016). Preprint, arXiv:1611.04076
35. Murez, Z., Kolouri, S., Kriegman, D., Ramamoorthi, R., Kim, K.: Image to image translation for domain adaptation. In: Proceedings of the IEEE Conference on Computer Vision and Pattern Recognition, pp. 4500–4509 (2018)
36. Netzer, Y., Wang, T., Coates, A., Bissacco, A., Wu, B., Ng, A.Y.: Reading digits in natural images with unsupervised feature learning. In: NIPS Workshop on Deep Learning and Unsupervised Feature Learning, vol. 2011, p. 5 (2011)
37. Noroozi, M., Vinjimoor, A., Favaro, P., Pirsiavash, H.: Boosting self-supervised learning via knowledge transfer. In: Proceedings of the IEEE Conference on Computer Vision and Pattern Recognition, pp. 9359–9367 (2018)
38. Ravi, S., Larochelle, H.: Optimization as a model for few-shot learning (2016)
39. Richter, S.R., Vineet, V., Roth, S., Koltun, V.: Playing for data: ground truth from computer games. In: European Conference on Computer Vision, pp. 102–118. Springer, Berlin (2016)
40. Ros, G., Sellart, L., Materzynska, J., Vazquez, D., Lopez, A.: The SYNTHIA Dataset: a large collection of synthetic images for semantic segmentation of urban scenes. In: CVPR (2016)
41. Rostami, M., Kolouri, S., Eaton, E., Kim, K.: Deep transfer learning for few-shot SAR image classification (2019)
42. Royer, A., Bousmalis, K., Gouws, S., Bertsch, F., Moressi, I., Cole, F., Murphy, K.: Xgan: unsupervised image-to-image translation for many-to-many mappings (2017). Preprint, arXiv:1711.05139
43. Saenko, K., Kulis, B., Fritz, M., Darrell, T.: Adapting visual category models to new domains. Computer Vision–ECCV 2010, pp. 213–226 (2010)
44. Sankaranarayanan, S., Balaji, Y., Castillo, C.D., Chellappa, R.: Generate to adapt: aligning domains using generative adversarial networks (2017). Preprint, arXiv:1704.01705
45. Sener, O., Song, H.O., Saxena, A., Savarese, S.: Learning transferrable representations for unsupervised domain adaptation. In: Advances in Neural Information Processing Systems, pp. 2110–2118 (2016)
46. Shrivastava, A., Pfister, T., Tuzel, O., Susskind, J., Wang, W., Webb, R.: Learning from simulated and unsupervised images through adversarial training. In: Proc. IEEE Conf. Computer Vision and Pattern Recognition (2017)
47. Snell, J., Swersky, K., Zemel, R.: Prototypical networks for few-shot learning. In: Advances in Neural Information Processing Systems, pp. 4077–4087 (2017)

48. Socher, R., Ganjoo, M., Manning, C.D., Ng, A.: Zero-shot learning through cross-modal transfer. In: Advances in Neural Information Processing Systems, pp. 935–943 (2013)
49. Srivastava, A., Valkov, L., Russell, C., Gutmann, M.U., Sutton, C.: Veegan: reducing mode collapse in gans using implicit variational learning. In: Guyon, I., Luxburg, U.V., Bengio, S., Wallach, H., Fergus, R., Vishwanathan, S., Garnett, R. (eds.) Advances in Neural Information Processing Systems, vol. 30, pp. 3308–3318. Curran Associates, Red Hook (2017). http://papers.nips.cc/paper/6923-veegan-reducing-mode-collapse-in-gans-using-implicit-variational-learning.pdf
50. Tabor, J., Knop, S., Spurek, P., Podolak, I., Mazur, M., Jastrzębski, S.: Cramer-wold autoencoder (2018). Preprint, arXiv:1805.09235
51. Taigman, Y., Polyak, A., Wolf, L.: Unsupervised cross-domain image generation (2016). Preprint, arXiv:1611.02200
52. Tsai, Y.H., Hung, W.C., Schulter, S., Sohn, K., Yang, M.H., Chandraker, M.: Learning to adapt structured output space for semantic segmentation. In: The IEEE Conference on Computer Vision and Pattern Recognition (CVPR) (2018)
53. Tzeng, E., Hoffman, J., Zhang, N., Saenko, K., Darrell, T.: Deep domain confusion: maximizing for domain invariance (2014). Preprint, arXiv:1412.3474
54. Tzeng, E., Hoffman, J., Darrell, T., Saenko, K.: Simultaneous deep transfer across domains and tasks. In: Proceedings of the IEEE International Conference on Computer Vision, pp. 4068–4076 (2015)
55. Tzeng, E., Hoffman, J., Saenko, K., Darrell, T.: Adversarial discriminative domain adaptation (2017). Preprint, arXiv:1702.05464
56. Venkateswara, H., Eusebio, J., Chakraborty, S., Panchanathan, S.: Deep hashing network for unsupervised domain adaptation. In: 2017 IEEE Conference on Computer Vision and Pattern Recognition (CVPR), pp. 5385–5394. IEEE, Piscataway (2017)
57. Yan, H., Ding, Y., Li, P., Wang, Q., Xu, Y., Zuo, W.: Mind the class weight bias: weighted maximum mean discrepancy for unsupervised domain adaptation. In: Proceedings of the IEEE Conference on Computer Vision and Pattern Recognition, pp. 2272–2281 (2017)
58. Yu, F., Koltun, V., Funkhouser, T.: Dilated residual networks (2017). Preprint, arXiv:1705.09914
59. Zhu, J.Y., Park, T., Isola, P., Efros, A.A.: Unpaired image-to-image translation using cycle-consistent adversarial networks (2017). Preprint, arXiv:1703.10593
60. Zou, Y., Yu, Z., Vijaya Kumar, B., Wang, J.: Unsupervised domain adaptation for semantic segmentation via class-balanced self-training. In: Proceedings of the European Conference on Computer Vision (ECCV), pp. 289–305 (2018)

Chapter 8
Domain Adaptation via Image Style Transfer

Amir Atapour-Abarghouei and Toby P. Breckon

8.1 Introduction

Recent advances in modern machine learning techniques have resulted in a significant growth in various computer vision applications readily deployed in real-world scenarios. However, the bias occasionally present within the datasets used to train these machine learning models can lead to notable issues. Such learning-based models often approximate functions capable of performing classification and prediction based tasks by capturing the underlying data distribution from which their training data is sampled. However, even small variations between the distributions of the training and the test data can negatively affect the performance of the approach. Such concerns have led to the creation of the field of transfer learning and domain adaptation [40], with a large community of researchers actively addressing the problem of data domain shift.

In this chapter, our primary focus is on the use of image style transfer as a domain adaptation technique. We utilise one of the fastest-growing and most challenging areas of research, namely monocular depth estimation, within computer vision as a means to demonstrate the efficacy of domain adaptation via image style transfer.

As 3D imagery has become more prevalent within computer vision, accurate and efficient depth estimation is now of paramount importance within many vision-based systems. While plausible depth estimation has been possible for many years using conventional strategies such as stereo correspondence [47], structure from motion [10, 14], depth from shading and light diffusion [1, 52, 58] and alike, such

A. Atapour-Abarghouei (✉)
Department of Computer Science, Durham University, Durham, UK
e-mail: amir.atapour-abarghouei@durham.ac.uk

T. P. Breckon
Departments of Engineering and Computer Science, Durham University, Durham, UK
e-mail: toby.breckon@durham.ac.uk

© Springer Nature Switzerland AG 2020
H. Venkateswara, S. Panchanathan (eds.), *Domain Adaptation in Computer Vision with Deep Learning*, https://doi.org/10.1007/978-3-030-45529-3_8

techniques often suffer from a myriad of issues such as intensive computational and calibration requirements, depth inhomogeneity and missing depth information, often resulting in the need for a post-processing stage to create more accurate and complete scene depth [2–6, 8, 35, 43]. Learning-based monocular depth estimation can offer a way to circumvent such issues as a novel alternative to many of these outdated approaches [6, 16, 19, 21, 31, 34, 59, 64].

Supervised learning-based monocular depth estimation approaches take advantage of off-line training on ground truth depth data to make depth prediction possible [16, 17, 31, 34, 67]. However, since ground truth depth is often scant and expensive to acquire in the real world, the practical use of many such approaches is heavily constrained.

There are, however, other monocular depth estimation approaches that do not require direct ground truth depth, but instead utilise a secondary supervisory signal during training which indirectly results in producing the desired depth [12, 19, 21, 59, 64]. Training data for these approaches is abundant and easily obtainable but they suffer from undesirable artefacts, such as blurring and incoherent content, due to the nature of their secondary supervision. However, an often overlooked fact is that the same technology that facilitates training large-scale deep neural networks can also assist in acquiring synthetic data for these neural networks [39, 49]. Nearly photo-realistic graphically rendered environments primarily used for gaming can be used to capture homogeneous synthetic depth images which are then utilised in training a depth estimating model.

While the use of such synthetic data is not novel and can resolve the issue of data scarcity [2, 18, 32, 49], the variations between synthetic and real-world images can lead to notable issues during deployment since any model trained on synthetic data cannot be expected to perform equally well when tested on naturally-sensed real-world images. Here, we intend to demonstrate the possibility of using style transfer as a domain adaptation technique. In this vein, in Sects. 8.2 and 8.3, we briefly outline the relevant areas of domain adaptation, image style transfer and their underlying connections and subsequently move on to practically demonstrating the applicability of style transfer in domain adaptation in the context of monocular depth estimation trained on synthetic imagery. This chapter is partially based on earlier work from [7].

8.2   Domain Adaptation via Maximum Mean Discrepancy

The main objective of domain adaptation is to transfer a model that has encapsulated the underlying distribution of a set of labelled data from the source domain so that it can perform well on previously-unseen unlabelled data from the target domain [40].

Within the current literature, this is often accomplished by minimising the distance between the source and target distributions. One of the most common metrics used to measure the distance between the two distributions is Maximum

Mean Discrepancy (MMD), which is the difference between probability measures based on embedding probabilities in a reproducing kernel Hilbert space [23, 53].

Assume there exist two sample sets $X = \{x_1, ..., x_n\}$ and $Y = \{y_1, ..., y_m\}$ with x_i and y_i independently and identically distributed from p and q respectively. As described in [23], in the two-sample testing problem, MMD can be used as a test statistic by drawing samples from distributions p and q and fitting a smooth function, which is large on points drawn from p and small on point drawn from q [23]. MMD is the difference between the mean function values on the two samples. This means when the samples are from different distributions, the MMD will be large and when the distributions are equal ($p = q$), the population MMD vanishes. More formally, the squared MMD is as follows [33]:

$$MMD^2[X, Y] = ||\mathbb{E}_x[\phi(x)] - \mathbb{E}_y[\phi(y)]||^2 = ||\frac{1}{n}\sum_{i=1}^{n}\phi(x_i) - \frac{1}{m}\sum_{j=1}^{m}\phi(y_j)||^2 =$$

$$\frac{1}{n^2}\sum_{i=1}^{n}\sum_{i'=1}^{n}\phi(x_i)^T\phi(x_i') + \frac{1}{m^2}\sum_{j=1}^{m}\sum_{j'=1}^{m}\phi(y_j)^T\phi(y_j') - \frac{2}{nm}\sum_{i=1}^{n}\sum_{j=1}^{m}\phi(x_i)^T\phi(y_j)$$

$$(8.1)$$

where $\phi(.)$ denotes the feature mapping function from X to $\mathbb{R}$. To reformulate Eq. (8.1) in the form of kernel, the function $k(x, y) = \langle\phi(x), \phi(y)\rangle_{\mathscr{H}}$ in a reproducing kernel Hilbert space $\mathscr{H}$ can be applied to the equation [33]:

$$MMD^2[X, Y] = \frac{1}{n^2}\sum_{i=1}^{n}\sum_{i'=1}^{n}k(x_i, x_i') + \frac{1}{m^2}\sum_{j=1}^{m}\sum_{j'=1}^{m}k(y_j, y_j')$$

$$-\frac{2}{nm}\sum_{i=1}^{n}\sum_{j=1}^{m}k(x_i, y_j)$$

$$(8.2)$$

where $k(., .)$ is the kernel function defining a mapping to a higher dimensional feature space. In Sect. 8.3, we provide a brief overview of the advances made in modern neural-based style transfer and its connections with domain adaptation via Maximum Mean Discrepancy.

8.3   Image Style Transfer

Image style transfer via convolutional neural networks first emerged as an effective stylization technique via the work in [20] and various improved and novel approaches capable of transferring the style of one image onto another [11, 29, 54] have been proposed ever since.

Conventionally, the style of an image is represented as a set of Gram matrices [48] that describe the correlations between low-level convolutional features extracted from the image, while the raw values of high-level semantic features often constitute the content of an image. These style and content representations are often extracted from a pre-trained loss network and are subsequently utilised to quantify style and content losses with respect to the target style and content images. More formally, the content loss for a specific layer l of the loss network can be defined as:

$$\mathcal{L}_{content} = \sum_{i=1}^{N_l} \sum_{j=1}^{M_l} ||f_{ij}^l(x) - f_{ij}^l(c)||^2 \qquad (8.3)$$

where c and x respectively denote the content and the output stylized images, f represents the loss network [51], $f^l(x)$ is the set of feature maps extracted from layer l after x is passed through f, N_l is the number of feature maps in layer l and M_l denotes the size (height $\times$ width) of the feature map. Similarly the style loss for a specific layer l of the loss network can be expressed as:

$$\mathcal{L}_{style} = \frac{1}{4N_l^2 M_l^2} \sum_{i=1}^{N_l} \sum_{j=1}^{N_l} (\mathcal{G}[f_{ij}^l(x)] - \mathcal{G}[f_{ij}^l(s)])^2 \qquad (8.4)$$

where s and x respectively represent the style and the output stylized images, and $\mathcal{G}[f_{ij}^l(x)]$ denotes the Gram matrix of the feature maps extracted from layer l after x is passed through f. The overall loss function can subsequently be defined as:

$$\mathcal{L} = \lambda_c \mathcal{L}_{content}(x, c) + \lambda_s \mathcal{L}_{style}(x, s) \qquad (8.5)$$

where λ_c and λ_s are coefficients determining the relative weights of the style and content loss components in the overall objective. In the original work in [20], this objective was minimised directly by gradient descent within the image space, and although the results of [20] are impressive, its process is very computationally intensive, leading to the emergence of alternative approaches that use neural networks to approximate the global minimum of the objective in a single forward pass. Such approaches [11, 29, 54] utilise neural networks trained to restyle an input image while preserving its content.

Style transfer can be considered as a distribution alignment process from the content image to the style image [28, 33]. In other words, transferring the style of one image (from the source domain) to another image (from the target domain) is essentially the same as minimising the distance between the source and target distributions. To demonstrate this connection between style transfer and domain adaptation (through MMD), the style loss in Eq. (8.4) can be reformulated by expanding the Gram matrices and applying the second order degree polynomial $k(x, y) = (x^T y)$ as follows [33]:

$$\mathcal{L}_{style} = \frac{1}{4N_l^2 M_l^2} \sum_{k_1=1}^{M_l} \sum_{k_2=1}^{M_l} (k(f_{k_1}^l(x), f_{k_2}^l(x)) + k(f_{k_1}^l(s), f_{k_2}^l(s))$$

$$- 2k(f_{k_1}^l(x), f_{k_2}^l(s))) = \frac{1}{4N_l^2} MMD^2[\mathcal{F}^l(x), \mathcal{F}^l(s)] \qquad (8.6)$$

where s and x respectively denote the style and the output stylized images, $f_k^l(x)$ denotes the kth column of $f^l(x)$ and $\mathcal{F}^l(x)$ is the set of features extracted from x in which each sample is a column of $f^l(x)$. Consequently, by minimising the style loss (reducing the distance between the style of the target image and desired stylized image), we are in effect reducing the distance between their distributions.

Here, we take advantage of this direct connection between domain adaptation and style transfer to perform monocular depth estimation by adapting our data distribution (i.e. real-world images) to our depth estimation model trained on data from a different distribution (i.e. synthetic images). However, while style transfer by matching Gram matrices is theoretically equivalent to minimising the MMD with the second order polynomial kernel and leads to domain adaptation, we forego the use of conventional style transfer and opt for an adversarially trained style transfer approach [66]. Other than the fact that the adversarially trained style transfer approach originally proposed in [66] is capable of superior performance and more pronounced changes in the style of the output image, the main reason for the choice of this approach is that [66] can transfer the style between two *sets* of unaligned images from different domains, while more conventional neural style transfer techniques such as [29] can only accept *one* specific image to be used as the style image. Within domain adaptation, this is not very desirable, especially since not one but tens of thousands of images representing the same style exist within the target domain. Experiments empirically justifying this choice are included in Sect. 8.5.2. In the next section, the approach to monocular depth estimation via style transfer [7] is outlined in greater depth.

8.4   Monocular Depth Estimation via Style Transfer

Our style transfer based monocular depth estimation approach consists of two stages, relying on two completely separate models trained at the same time to carry out the operations of each stage. The first stage includes directly training a depth estimation model using synthetic data captured from a graphically rendered environment primarily designed for gaming applications [39] (Sect. 8.4.1). However, as the eventual objective of the overall model involves estimating depth from real-world images, we attempt to reduce the domain discrepancy between the synthetic data distribution and the real-world data distribution using a model trained to transfer the style between synthetic images and real-world images in the second stage of the overall approach (Sect. 8.4.2).

8.4.1 Stage 1: Depth Estimation Model

Here, we consider monocular depth estimation as an image-to-image mapping problem, with the RGB image used as the input to our mapping function, and scene depth produced as its output. Using modern convolutional neural networks, image-to-image translation and prediction problems have become significantly more tractable and can yield remarkably high-quality results. An overly simplistic solution to a translation problem such as depth estimation would be employing a network that attempts to minimise a reconstruction loss (Euclidean distance) between the pixel values of the output and the ground truth. However, since monocular depth estimation is an inherently multi-modal problem (a problem that has several global solutions instead of a unique global optimum since several plausible depth values can correspond with a single RGB view), any model trained to predict depth based on a sole reconstruction loss tends to generate values that are the average of all the possible modes in the predictions. This averaging can lead to blurring effects in the outputs.

As a result, many such prediction-based approaches [2, 7, 27, 42, 60, 61, 63, 66] and other generative models [15, 55] make use of adversarial training [22] to alleviate the blurry output problem since the use of an adversarial loss generally forces the model to select a single mode from the distribution instead of averaging all possible modes and generate more realistic results without blurring.

A Generative Adversarial Network (GAN) [22] is capable of producing semantically sound samples by creating a competition between a generator, which attempts to capture the underlying data distribution, and a discriminator, which judges the output of the generator and penalises unrealistic images and artefacts. Both networks are trained simultaneously to achieve an equilibrium [22]. Whilst most generative models generate images from a latent noise vector as the input to the generator, the model presented here is solely conditioned on an input image (RGB).

More formally, the generative model learns a mapping from the input image, x (RGB view), to the output image, y (scene depth), $G : x \rightarrow y$. The generator, G, attempts to produce fake samples, $G(x) = \tilde{y}$, which cannot be distinguished from real ground truth samples, y, by the discriminator, D, which is adversarially trained to detect the fake samples produced by the generator.

Many other approaches following a similar framework incorporate a random noise vector z or drop-outs into the generator training to prevent deterministic mapping and induce stochasticity [27, 37, 42, 57]. However, since deterministic mapping is not of concern in a problem such as depth estimation, no random noise or drop-out is required. Empirical experiments demonstrate no significant difference in the output distribution could be achieved even if stochasticity is encouraged within the model using these strategies.

8.4.1.1 Loss Function

The objective of the monocular depth estimation model is achieved via minimising a loss function consisting of two components. The first is a simple reconstruction loss, which forces the generator to capture the structural and contextual content of the scene and output depth images which are as close as possible to the ground truth depth information. To accomplish this, we use the L_1 loss:

$$\mathcal{L}_{rec} = ||G(x) - y||_1 \tag{8.7}$$

While the use of a reconstruction loss can help the network to internally model the structure and content of the scene, it can also lead to the generator optimising towards averaging all possible output depth values rather than selecting one, which can lead to blurring effects within the output depth image. Consequently, the second component of the overall loss function, an adversarial loss, is introduced to incentivise the generator to create shaper and higher quality depth images:

$$\mathcal{L}_{adv} = \min_{G} \max_{D} \mathbb{E}_{x,y\sim\mathbb{P}_d(x,y)} [\log D(x, y)] + \mathbb{E}_{x\sim\mathbb{P}_d(x)} [\log(1 - D(x, G(x)))] \tag{8.8}$$

where $\mathbb{P}_d$ denotes the data distribution defined by $\tilde{y} = G(x)$ and x is the input to the generator and y the ground truth. Subsequently, the overall loss function is as follows:

$$\mathcal{L} = \lambda\mathcal{L}_{rec} + (1 - \lambda)\mathcal{L}_{adv} \tag{8.9}$$

with λ being the weighting coefficient selected empirically.

8.4.1.2 Implementation Details

In order to obtain the synthetic data required to train the depth estimation model, corresponding colour and disparity images are captured using a camera view placed in front of a virtual car as it automatically drives around a graphically-rendered virtual environment, with images captured every 60 frames with randomly varying height, field of view, weather and lighting conditions at different times of day. From the overall dataset of 80,000 corresponding pairs of colour and disparity images captured in this manner, 70,000 are used for training and 10,000 are set aside for testing. The depth estimation model trained using the synthetic dataset generates disparity images which can be converted to depth using the known camera parameters and scaled to the depth range of the KITTI image frame [38].

An important aspect of any depth estimation problem is that the overall structure and the high frequency information present within the RGB view of the scene (input) and the depth image (output) are aligned as they ultimately represent the exact same scene. As a result, much information (e.g. structure, geometry, object boundaries and alike) is shared between the input and output. Consequently, we utilise the capabilities of skip connections within the architecture of the generator [9, 25, 42, 45, 57] to accurately preserve high-frequency scene content. The generator, therefore, can take advantage of the opportunity to directly pass geometric information between corresponding layers in the encoder and the decoder without having to go through every single layer in between and possibly losing precious details in the down-sampling and up-sampling processes.

The generator consists of an architecture similar to that of [45] with the exception that skip connections exist between every pair of corresponding layers in the encoder and decoder. As for the discriminator, the basic architecture used in [44] is deployed. Both the generator and discriminator utilise convolution-BatchNorm-ReLu modules [26] with the discriminator using leaky ReLUs (*slope* = 0.2).

All technical implementation is performed using *PyTorch* [41], with Adam [30] providing the optimisation ($\beta_1 = 0.5$, $\beta_2 = 0.999$, $\alpha = 0.0002$). The weighting coefficient in the overall loss function in Eq. (8.9) was empirically chosen to be $\lambda = 0.99$.

8.4.2 Stage 2: Style Transfer as Domain Adaptation

The monocular depth estimation model presented in Sect. 8.4.1 can perform very well on unseen images from the test set of synthetic data captured from the virtual environment. However, since the model is only trained on synthetic images and the synthetic and real-world images are from different domains, directly estimating depth from RGB images captured in the real-world remains challenging, which is why domain adaptation via style transfer is an important component of the overall approach.

The objective of the style transfer component of the approach, therefore, is to learn a mapping function $\mathcal{D} : X \rightarrow Y$ from the source domain X (real-world images) to the target domain Y (synthetic images) in a way that the distributions $\mathcal{D}(X)$ and Y are identical. When images from X are mapped into Y, their corresponding depth information can be inferred using the monocular depth estimation model presented in Sect. 8.4.1 that is specifically trained on images from Y.

Within the existing literature, there have been various successful attempts at transforming images from one domain to another [36, 46, 50, 66]. Here, the proposed approach relies on the idea of image style transfer using generative adversarial networks, as proposed in [66], to reduce the discrepancy between the source domain (real-world data) and the target domain (synthetic data on which the depth estimation model in Sect. 8.4.2 trained). This approach uses adversarial training [22] and cycle-consistency [21, 56, 62, 65] to translate between two sets of unaligned images from different domains.

More formally, the objective is to map images between the two domains X and Y with the respective distributions of $x \sim \mathbb{P}_d(x)$ and $y \sim \mathbb{P}_d(y)$. The mapping

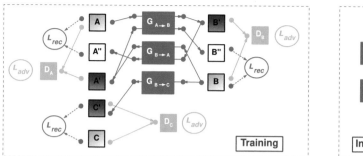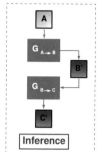

Fig. 8.1 Outline of the approach to monocular depth estimation via domain adaptation using [66]. Domain A (real-world RGB) is transformed into B (synthetic RGB) and then to C (pixel-perfect depth). A, B and C represent ground truth images, A', B' and C' are the generated images and A" and B" denote images cyclically regenerated via [66]

functions are approximated using two separate generators, G_{XtoY} and G_{YtoX} and two discriminators D_X (discriminating between $x \in X$ and $G_{YtoX}(y)$) and D_Y (discriminating between $y \in Y$ and $G_{XtoY}(x)$). The loss contains two components: an adversarial loss [22] and a cycle consistency loss [66]. The general pipeline of the approach (along with the depth estimation model Sect. 8.4.1) is seen in Fig. 8.1, with three generators $G_{A_{to}B}$, $G_{B_{to}A}$ and $G_{B_{to}C}$, and three discriminators D_A, D_B and D_C.

8.4.2.1 Loss Function

Since there are two generators to constrain the content of the images, there are two mapping functions. The use of an adversarial loss guarantees the style of one domain is transferred to the other. The loss for G_{XtoY} with D_Y is represented as follows:

$$\mathcal{L}_{adv-XtoY} = \min_{G_{XtoY}} \max_{D_Y} \mathop{\mathbb{E}}_{y\sim\mathbb{P}_d(y)} [\log D_Y(y)] + \mathop{\mathbb{E}}_{x\sim\mathbb{P}_d(x)} [\log(1 - D_Y(G_{XtoY}(x)))]$$

(8.10)

where $\mathbb{P}_d$ is the data distribution, X the source domain with samples x and Y the target domain with samples y. Similarly, for G_{YtoX} and D_X, the adversarial loss is as follows:

$$\mathcal{L}_{adv-YtoX} = \min_{G_{YtoX}} \max_{D_X} \mathop{\mathbb{E}}_{x\sim\mathbb{P}_d(x)} [\log D_X(x)] + \mathop{\mathbb{E}}_{y\sim\mathbb{P}_d(y)} [\log(1 - D_X(G_{YtoX}(y)))]$$

(8.11)

To constrain the adversarial loss of the generators to force the model to produce contextually coherent images rather than random semantically meaningless content from the target domain, a cycle-consistency loss is added that encourages the model to become capable of bringing an image x that is translated into the target domain Y using G_{XtoY} back into the source domain X using G_{YtoX}. In essence, after a full cycle: $G_{YtoX}(G_{XtoY}(x)) = x$ and vice versa. Consequently, the cycle-consistency loss is as follows:

$$\mathcal{L}_{cyc} = ||G_{YtoX}(G_{XtoY}(x)) - x||_1 + ||G_{XtoY}(G_{YtoX}(y)) - y||_1 \qquad (8.12)$$

Subsequently, the joint loss function is as follows:

$$\mathcal{L} = \mathcal{L}_{adv-XtoY} + \mathcal{L}_{adv-YtoX} + \lambda\mathcal{L}_{cyc} \qquad (8.13)$$

with λ being the weighting coefficient selected empirically.

8.4.2.2   Implementation Details

The architecture of the generators is similar to that of the network proposed in [29] with two convolutional layers followed by nine residual blocks [24] and two up-convolutions that bring the image back to its original input size. As for the discriminators, the same architecture is used as was in Sect. 8.4.1. Moreover, the discriminators are updated based on the last 50 generator outputs and not just the last generated image [50, 66].

All technical implementation is performed using *PyTorch* [41], with Adam [30] providing the optimisation ($\beta_1 = 0.5$, $\beta_2 = 0.999$, $\alpha = 0.0001$). The weighting coefficient in the overall loss function in Eq. (8.13) was empirically chosen to be $\lambda = 10$.

8.5   Experimental Results

In order to demonstrate the efficacy of style transfer used as domain adaptation technique for the task of monocular depth estimation, in this section, the depth estimation approach is evaluated using ablation studies and both qualitative and quantitative comparisons with state-of-the-art monocular depth estimation methods. The KITTI dataset [38] and locally-captured data are used for evaluations.

8.5.1   Comparisons Against Contemporary Approaches

To evaluate the performance of the monocular depth estimation approach in Sect. 8.4 and demonstrate the capability of style transfer used as domain adaptation, 697 images from the data split suggested in [17] are used as the test set. As demonstrated in Table 8.1, the monocular depth estimation model trained on synthetic data and adapted using style transfer (DST) performs better than contemporary monocular depth estimation approaches directly trained on real-world images [17, 21, 34, 64]

Table 8.1 Comparing the results of depth estimation via style transfer (DST) against other approaches over the KITTI dataset using the data split in [17]

Method	Training data	Error metrics (lower, better)				Accuracy metrics (higher, better)		
		Abs. Rel.	Sq. Rel.	RMSE	RMSE log	$\sigma < 1.25$	$\sigma < 1.25^2$	$\sigma < 1.25^3$
Eigen et al. coarse	K	0.214	1.605	6.563	0.292	0.673	0.884	0.957
Eigen et al. fine	K	0.203	1.548	6.307	0.282	0.702	0.890	0.958
Liu et al.	K	0.202	1.614	6.523	0.275	0.678	0.895	0.965
Zhou et al.	K	0.208	1.768	6.856	0.283	0.678	0.885	0.957
Zhou et al.	K+CS	0.198	1.836	6.565	0.275	0.718	0.901	0.960
Godard et al.	K	0.148	1.344	5.927	0.247	0.803	0.922	0.964
Godard et al.	K+CS	0.124	1.076	5.311	0.219	0.847	0.942	0.973
DST approach	K+S*	**0.110**	**0.929**	**4.726**	**0.194**	**0.923**	**0.967**	**0.984**

For the training data, K represents KITTI [38], CS Cityscapes [13] and S* the synthetic data captured from a virtual environment
Bold values indicate the best results

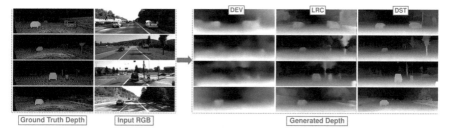

Ground Truth Depth   Input RGB   Generated Depth

Fig. 8.2 Qualitative comparison of the results of contemporary state-of-the-art approaches of depth and ego-motion from video (DEV) [64], estimation based on left/right consistency (LRC) [21] and depth via style transfer (DST) over the KITTI split

with lower error and higher accuracy. Some of the comparators [21, 64] use a combination of different datasets for training and fine-tuning to boost performance, while the approach presented here only relies on synthetic data for training.

The data split of 200 images in KITTI [38] is also used to provide better qualitative evaluation, since the ground truth disparity images within this split are of considerably higher quality and provide CAD models as replacements for moving cars. As seen in Fig. 8.2, compared to other approaches [21, 64] trained on similar data domains, monocular depth estimation via style transfer leads to sharper and more crisp outputs in which object boundaries and thin structures are better preserved.

8.5.2 Ablation Studies

Ablation studies are integral in demonstrating the necessity of the components of the approach. The monocular depth estimation model presented in Sect. 8.4.1

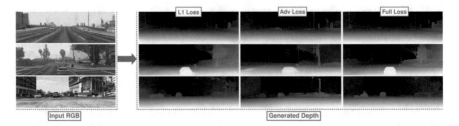

Fig. 8.3 Demonstrating the importance of the different components of the loss function in the depth estimation model (Sect. 8.4.1)

utilises a combination of reconstruction and adversarial losses (Eq. (8.9)). In order to test the importance of each loss component, the model is separately trained using the reconstruction loss only and the adversarial loss only. Figure 8.3 demonstrates the effects of removing parts of the training objective. The model based only on the reconstruction loss produces contextually sound but blurry results, while the adversarial loss generates sharp outputs that contain artefacts. When the approach is trained using the full overall loss function, it creates more accurate results without unwanted effects. Further numerical and qualitative evidence of the efficacy of a combination of a reconstruction and adversarial loss can be found in [27].

Another important aspect of the ablation studies involves demonstrating the necessity of domain adaptation (Sect. 8.4.2) within the overall pipeline. As indicated in Table 8.2, due to the differences in the domains of the synthetic and natural data, the depth estimation model directly applied to real-world data does not produce numerically desirable results, which points to the importance of domain adaptation to the approach. Similarly, Fig. 8.4 qualitatively demonstrates that when no style transfer is used in the approach, the generated depth outputs contain significant inaccuracies and undesirable artefacts.

While the connection between domain adaptation by minimising the Maximum Mean Discrepancy with the second order polynomial kernel and neural style transfer by matching Gram matrices is briefly outlined in Sects. 8.2 and 8.3, the approach presented here does not use conventional neural style transfer and instead requires an adversarial discriminator [66] to carry out style transfer for domain adaptation.

To demonstrate that a discriminator can reasonably perform domain adaptation via style transfer, experiments are carried out with the style transfer approach proposed in [29], which improves on the pioneering style transfer work of [20] by training a generator that can transfer a specific style (that of our synthetic domain in this work) onto a set of images of a specific domain (real-world images) by minimising content and style losses (Eqs. (8.3) and (8.4)). An overview of the entire pipeline using [29] (along with the monocular depth estimation model in Sect. 8.4.1) is seen in Fig. 8.5.

Whilst [66] is capable of transferring the style between two large *sets* of unaligned images from different domains, the neural style transfer approach in [29] requires *one* specific image to be used as the target style image. In this work,

Table 8.2 Ablation study over the KITTI dataset using the KITTI split

Method	Training data	Error metrics (lower, better)				Accuracy metrics (higher, better)		
		Abs. Rel.	Sq. Rel.	RMSE	RMSE log	$\sigma < 1.25$	$\sigma < 1.25^2$	$\sigma < 1.25^3$
w/o domain adaptation	K+S*	0.498	6.533	9.382	0.609	0.712	0.823	0.883
w/ the approach in [29]	K+S*	0.154	1.338	6.470	0.296	0.874	0.962	0.981
w/ the approach in [66]	K+S*	**0.101**	**1.048**	**5.308**	**0.184**	**0.903**	**0.988**	**0.992**

The approach is trained using, KITTI (K) and synthetic data (S*). The approach provides the best results when it includes domain adaptation via style transfer using the technique in [66]

Bold values indicate the best results

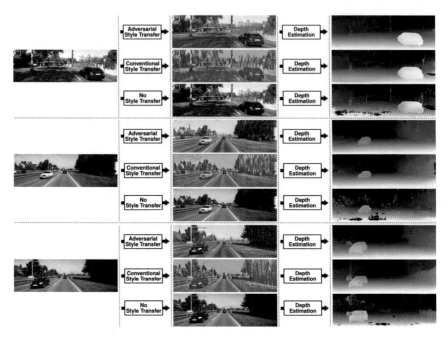

Fig. 8.4 Exemplar results demonstrating the importance of style transfer. Examples include results of depth estimation with style transfer via cycle-consistent adversarial training [66], conventional style transfer approach of [29] and without any style transfer

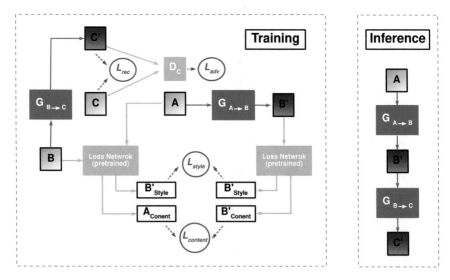

Fig. 8.5 Outline of the approach to monocular depth estimation via domain adaptation using [29]. Images from domain A (real-world) are transformed into B (synthetic) and then to C (pixel-perfect depth). A, B and C denote ground truth images and A', B' and C' represent generated images

the target domain consists of tens of thousands of images representing the same style. Consequently, a number of synthetic images that contain a variety of objects, textures and colours that represent their domain are collected and a single image that holds the desired style is created by pooling features from the images.

To evaluate the performance of the approach regarding the effects of domain adaptation via style transfer, the data split of 200 images in the KITTI dataset [38] is used. Experiments are carried out with both style transfer techniques in [66] and [29], in addition to using real-world images as direct inputs to the depth estimation model without any domain adaptation. As seen in the results presented in Table 8.2, using direct real-world inputs without any domain adaptation via style transfer results in significant anomalies in the output while translating images into synthetic space using [66] before depth estimation leads to notably improved results. The qualitative results provided in Fig. 8.4 also point to the same conclusion.

8.5.3 Generalisation

The use of domain adaptation via style transfer can make the model more robust and less susceptible to domain shift in the presence of unseen data. Considering that the images used in the training procedure of the monocular depth estimation model (Sect. 8.4.1) are captured from a synthetic environment [39] and the data used to train the style transfer component of the approach (Sect. 8.4.2) are from the KITTI dataset [38], we evaluate the generalisation capabilities of the approach using additional data captured locally in an urban environment. As clearly seen in Fig. 8.6, the approach is easily capable of generating sharp, coherent and visually plausible depth without any training on the unseen images from the new data domain.

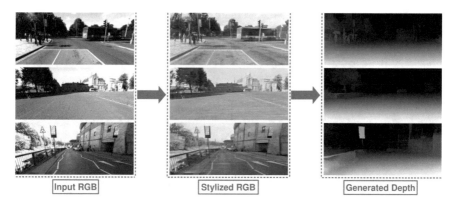

Input RGB   Stylized RGB   Generated Depth

Fig. 8.6 Qualitative results of the proposed approach on urban driving scenes captured locally without further training

8.6 Limitations

The monocular depth estimation approach discussed in this chapter is capable of generating high quality and accurate depth with minimal anomalies by taking advantage of domain adaptation via image style transfer. However, the very component of the approach that enables it to generate highly accurate pixel-perfect depth, namely style transfer, can also bring forth certain shortcomings within the overall pipeline. The most significant issue is that of adapting to sudden lighting changes and saturation during style transfer. The two domains of images used here (synthetic and real-world images) significantly vary in intensity differences between lit areas and shadows, as is very common in images captured using different cameras in different environments. As a result, image regions containing shadows can be wrongly construed as elevated surfaces or foreground objects post style transfer, leading to inaccurate depth estimation of said regions. Examples in Fig. 8.7 demonstrate how such issues can arise.

Moreover, despite the fact that holes (missing regions) are generally considered undesirable in depth images [3, 4, 8, 35, 43], certain areas within the scene depth should remain without depth values (e.g. very distant objects and sky). However, a supervised monocular depth estimation approach such as the one discussed in this chapter is incapable of distinguishing the sky from other extremely saturated objects within the scene even with style transfer, which can lead to creation of small holes where they do not belong.

Additionally, while the approach in [66] has been demonstrated to be very powerful in mapping between two sets of unaligned images with similar content, it can be very susceptible to wrongly synthesising meaningless content. This can especially happen if certain content (scene objects, geometry, structure and alike) is commonly found in images from one domain but not the other. Under these circumstances, the adversarial discriminator in [66] tends to encourage the generator to synthesise content in the latter domain to compensate for the discrepancies induced by the differences in overall scene content. Since in domain adaptation via style transfer, the objective is to transform the style of the images and not their content, this issue can lead to significant issues in terms of unwanted artefacts and anomalies within the output.

Fig. 8.7 Examples of failures, mainly due to light saturation and shadows

8.7   Conclusion

In this chapter, we have primarily focused on demonstrating the viability of image style transfer as a domain adaptation technique in computer vision applications. The aim of a domain adaptation approach is to adapt a model trained on one set of data from a specific domain to perform well on previously-unseen data from a different domain. In this vein, we have selected the problem of monocular depth estimation for our experiments since large quantities of ground truth depth data required for training a directly supervised monocular depth estimation approach is extremely expensive and difficult to obtain, leading to a greater need for domain adaptation.

Taking advantage of pixel-perfect synthetic depth data captured from a graphically rendered urban environment designed for gaming applications, an effective depth estimation model can be trained in a directly supervised manner. However, such a model cannot be expected to perform well on previously-unseen real-world images as the data distributions to which images from these two domains (synthetic images and real-world images) belong are vastly different. Since modern advances in neural style transfer can theoretically be linked to minimising the Maximum Mean Discrepancy between two distributions with the second order polynomial kernel, we make use of an adversarially trained cycle-consistent approach capable of transferring styles between two unaligned sets of images to adapt our real-world data to fit into the distribution approximated by the generator in our depth estimation model. Despite certain isolated issues, experimental evaluations and comparisons against contemporary monocular depth estimation approaches demonstrate that style transfer is indeed a highly effective method of domain adaptation.

References

1. Abrams, A., Hawley, C., Pless, R.: Heliometric stereo: shape from sun position. In: Proc. Euro. Conf. Computer Vision, pp. 357–370 (2012)
2. Atapour-Abarghouei, A., Akcay, S., Payen de La Garanderie, G., Breckon, T.: Generative adversarial framework for depth filling via Wasserstein metric, cosine transform and domain transfer. Pattern Recognit. **91**, 232–244 (2019)
3. Atapour-Abarghouei, A., Breckon, T.: DepthComp: real-time depth image completion based on prior semantic scene segmentation. In: Proc. British Machine Vision Conference (2017)
4. Atapour-Abarghouei, A., Breckon, T.: A comparative review of plausible hole filling strategies in the context of scene depth image completion. Comput. Graphics **72**, 39–58 (2018)
5. Atapour-Abarghouei, A., Breckon, T.: Extended patch prioritization for depth filling within constrained exemplar-based RGB-D image completion. In: Proc. Int. Conf. Image Analysis and Recognition, pp. 306–314 (2018)
6. Atapour-Abarghouei, A., Breckon, T.: Veritatem Dies Aperit-temporally consistent depth prediction enabled by a multi-task geometric and semantic scene understanding approach (2019). Preprint, arXiv:1903.10764
7. Atapour-Abarghouei, A., Breckon, T.P.: Real-time monocular depth estimation using synthetic data with domain adaptation via image style transfer. In: Proc. IEEE Conf. Computer Vision and Pattern Recognition, pp. 2800–2810 (2018)

8. Atapour-Abarghouei, A., Payen de La Garanderie, G., Breckon, T.: Back to Butterworth - a Fourier basis for 3D surface relief hole filling within RGB-D imagery. In: Proc. Int. Conf. Pattern Recognition, pp. 2813–2818. IEEE, Piscataway (2016)
9. Badrinarayanan, V., Kendall, A., Cipolla, R.: SegNet: a deep convolutional encoder-decoder architecture for image segmentation. IEEE Trans. Pattern Anal. Mach. Intell. **39**(12), 2481–2495 (2017)
10. Cavestany, P., Rodriguez, A., Martinez-Barbera, H., Breckon, T.: Improved 3d sparse maps for high-performance structure from motion with low-cost omnidirectional robots. In: Proc. Int. Conf. Image Processing, pp. 4927–4931 (2015)
11. Chen, T.Q., Schmidt, M.: Fast patch-based style transfer of arbitrary style. In: Workshop in Constructive Machine Learning, pp. 1–5 (2016)
12. Chen, W., Fu, Z., Yang, D., Deng, J.: Single-image depth perception in the wild. In: Advances in Neural Information Processing Systems, pp. 730–738 (2016)
13. Cordts, M., Omran, M., Ramos, S., Rehfeld, T., Enzweiler, M., Benenson, R., Franke, U., Roth, S., Schiele, B.: The Cityscapes dataset for semantic urban scene understanding. In: Proc. IEEE Conf. Computer Vision and Pattern Recognition, pp. 3213–3223 (2016)
14. Ding, L., Sharma, G.: Fusing structure from motion and LiDAR for dense accurate depth map estimation. In: Proc. Int. Conf. Acoustics, Speech and Signal Processing, pp. 1283–1287. IEEE, Piscataway (2017)
15. Dosovitskiy, A., Brox, T.: Generating images with perceptual similarity metrics based on deep networks. In: Advances in Neural Information Processing Systems, pp. 658–666 (2016)
16. Eigen, D., Fergus, R.: Predicting depth, surface normals and semantic labels with a common multi-scale convolutional architecture. In: Proc. Int. Conf. Computer Vision, pp. 2650–2658 (2015)
17. Eigen, D., Puhrsch, C., Fergus, R.: Depth map prediction from a single image using a multi-scale deep network. In: Advances in Neural Information Processing Systems, pp. 2366–2374 (2014)
18. Gaidon, A., Wang, Q., Cabon, Y., Vig, E.: Virtual worlds as proxy for multi-object tracking analysis. In: Proc. IEEE Conf. Computer Vision and Pattern Recognition, pp. 4340–4349 (2016)
19. Garg, R., Carneiro, G., Reid, I.: Unsupervised CNN for single view depth estimation: geometry to the rescue. In: Proc. Euro. Conf. Computer Vision, pp. 740–756. Springer, Berlin (2016)
20. Gatys, L.A., Ecker, A.S., Bethge, M.: Image style transfer using convolutional neural networks. In: Proc. IEEE Conf. Computer Vision and Pattern Recognition, pp. 2414–2423 (2016)
21. Godard, C., Mac Aodha, O., Brostow, G.J.: Unsupervised monocular depth estimation with left-right consistency. In: Proc. IEEE Conf. Computer Vision and Pattern Recognition, pp. 6602–6611 (2017)
22. Goodfellow, I., Pouget-Abadie, J., Mirza, M., Xu, B., Warde-Farley, D., Ozair, S., Courville, A., Bengio, Y.: Generative adversarial nets. In: Advances in Neural Information Processing Systems, pp. 2672–2680 (2014)
23. Gretton, A., Borgwardt, K.M., Rasch, M.J., Schölkopf, B., Smola, A.: A kernel two-sample test. Mach. Learn. Res. **13**, 723–773 (2012)
24. He, K., Zhang, X., Ren, S., Sun, J.: Deep residual learning for image recognition. In: Proc. Conf. Computer Vision and Pattern Recognition, pp. 770–778 (2016)
25. Hinton, G.E., Salakhutdinov, R.R.: Reducing the dimensionality of data with neural networks. Science **313**(5786), 504–507 (2006)
26. Ioffe, S., Szegedy, C.: Batch normalization: accelerating deep network training by reducing internal covariate shift. In: Proc. Int. Conf. Machine Learning, pp. 448–456 (2015)
27. Isola, P., Zhu, J.Y., Zhou, T., Efros, A.A.: Image-to-image translation with conditional adversarial networks. In: Proc. IEEE Conf. Computer Vision and Pattern Recognition, pp. 5967–5976 (2017)
28. Jing, Y., Yang, Y., Feng, Z., Ye, J., Song, M.: Neural style transfer: a review (2017). Preprint, arXiv:1705.04058

29. Johnson, J., Alahi, A., Fei-Fei, L.: Perceptual losses for real-time style transfer and super-resolution. In: Proc. Euro. Conf. Computer Vision, pp. 694–711 (2016)
30. Kingma, D., Ba, J.: Adam: a method for stochastic optimization. In: Proc. Int. Conf. Learning Representations (2014)
31. Ladicky, L., Shi, J., Pollefeys, M.: Pulling things out of perspective. In: Proc. Conf. Computer Vision and Pattern Recognition, pp. 89–96 (2014)
32. Le, T.A., Baydin, A.G., Zinkov, R., Wood, F.: Using synthetic data to train neural networks is model-based reasoning. In: Proc. Int. Joint Conf. Neural Networks, pp. 3514–3521. IEEE, Piscataway (2017)
33. Li, Y., Wang, N., Liu, J., Hou, X.: Demystifying neural style transfer. In: Proc. Int. Joint Conf. Artificial Intelligence, pp. 2230–2236. AAAI Press, Palo Alto (2017)
34. Liu, F., Shen, C., Lin, G., Reid, I.: Learning depth from single monocular images using deep convolutional neural fields. IEEE Trans. Pattern Anal. Mach. Intell. **38**(10), 2024–2039 (2016)
35. Liu, J., Gong, X., Liu, J.: Guided inpainting and filtering for kinect depth maps. In: Proc. Int. Conf. Pattern Recognition, pp. 2055–2058. IEEE, Piscataway (2012)
36. Liu, M.Y., Tuzel, O.: Coupled generative adversarial networks. In: Advances in Neural Information Processing Systems, pp. 469–477 (2016)
37. Mathieu, M., Couprie, C., LeCun, Y.: Deep multi-scale video prediction beyond mean square error. In: Proc. Int. Conf. Learning Representations (2016)
38. Menze, M., Geiger, A.: Object scene flow for autonomous vehicles. In: Proc. IEEE Conf. Computer Vision and Pattern Recognition, pp. 3061–3070 (2015)
39. Miralles, R.: An open-source development environment for self-driving vehicles. In: Universitat Oberta de Catalunya, pp. 1–31 (2017)
40. Pan, S.J., Yang, Q.: A survey on transfer learning. IEEE Trans. Knowl. Data Eng. **22**(10), 1345–1359 (2010)
41. Paszke, A., Gross, S., Chintala, S., Chanan, G., Yang, E., DeVito, Z., Lin, Z., Desmaison, A., Antiga, L., Lerer, A.: Automatic differentiation in PyTorch. In: Advances in Neural Information Processing Systems, pp. 1–4 (2017)
42. Pathak, D., Krahenbuhl, P., Donahue, J., Darrell, T., Efros, A.A.: Context encoders: feature learning by inpainting. In: Proc. IEEE Conf. Computer Vision and Pattern Recognition, pp. 2536–2544 (2016)
43. Qi, F., Han, J., Wang, P., Shi, G., Li, F.: Structure guided fusion for depth map inpainting. Pattern Recognit. Lett. **34**(1), 70–76 (2013)
44. Radford, A., Metz, L., Chintala, S.: Unsupervised representation learning with deep convolutional generative adversarial networks, pp. 1–16 (2015). Preprint, arXiv:1511.06434
45. Ronneberger, O., Fischer, P., Brox, T.: U-net: convolutional networks for biomedical image segmentation. In: Proc. Int. Conf. Medical Image Computing and Computer-Assisted Intervention, pp. 234–241. Springer, Berlin (2015)
46. Rosales, R., Achan, K., Frey, B.J.: Unsupervised image translation. In: Proc. Int. Conf. Computer Vision, pp. 472–478 (2003)
47. Scharstein, D., Szeliski, R.: A taxonomy and evaluation of dense two-frame stereo correspondence algorithms. Int. J. Comput. Vis. **47**(1–3), 7–42 (2002)
48. Schwerdtfeger, H.: Introduction to Linear Algebra and the Theory of Matrices. P. Noordhoff, Groningen (1950)
49. Shah, S., Dey, D., Lovett, C., Kapoor, A.: AirSim: high-fidelity visual and physical simulation for autonomous vehicles. In: Field and Service Robotics, pp. 621–635 (2017)
50. Shrivastava, A., Pfister, T., Tuzel, O., Susskind, J., Wang, W., Webb, R.: Learning from simulated and unsupervised images through adversarial training. In: Proc. IEEE Conf. Computer Vision and Pattern Recognition, pp. 2242–2251 (2017)
51. Simonyan, K., Zisserman, A.: Very deep convolutional networks for large-scale image recognition. In: Proc. Int. Conf. Learning Representations, pp. 1–14 (2015)
52. Tao, M.W., Srinivasan, P.P., Malik, J., Rusinkiewicz, S., Ramamoorthi, R.: Depth from shading, defocus, and correspondence using light-field angular coherence. In: Proc. IEEE Conf. Computer Vision and Pattern Recognition, pp. 1940–1948 (2015)

53. Tolstikhin, I.O., Sriperumbudur, B.K., Schölkopf, B.: Minimax estimation of Maximum Mean Discrepancy with radial kernels. In: Advances in Neural Information Processing Systems, pp. 1930–1938 (2016)
54. Ulyanov, D., Lebedev, V., Vedaldi, A., Lempitsky, V.S.: Texture networks: feed-forward synthesis of textures and stylized images. In: Proc. Int. Conf. Machine Learning, pp. 1349–1357 (2016)
55. Walker, J., Marino, K., Gupta, A., Hebert, M.: The pose knows: video forecasting by generating pose futures. In: Proc. Int. Conf. Computer Vision, pp. 3352–3361. IEEE, Piscataway (2017)
56. Wang, F., Huang, Q., Guibas, L.J.: Image co-segmentation via consistent functional maps. In: Proc. Int. Conf. Computer Vision, pp. 849–856 (2013)
57. Wang, X., Gupta, A.: Generative image modeling using style and structure adversarial networks. In: Proc. Euro. Conf. Computer Vision, pp. 318–335. Springer, Berlin (2016)
58. Woodham, R.J.: Photometric method for determining surface orientation from multiple images. Opt. Eng. **19**(1), 191139 (1980)
59. Xie, J., Girshick, R., Farhadi, A.: Deep3D: fully automatic 2D-to-3D video conversion with deep convolutional neural networks. In: Proc. Euro. Conf. Computer Vision, pp. 842–857. Springer, Berlin (2016)
60. Yang, C., Lu, X., Lin, Z., Shechtman, E., Wang, O., Li, H.: High-resolution image inpainting using multi-scale neural patch synthesis. In: Proc. IEEE Conf. Computer Vision and Pattern Recognition, pp. 4076–4084 (2017)
61. Yeh, R.A., Chen, C., Lim, T.Y., Schwing, A.G., Hasegawa-Johnson, M., Do, M.N.: Semantic image inpainting with deep generative models. In: Proc. IEEE Conf. Computer Vision and Pattern Recognition, pp. 6882–6890 (2017)
62. Yi, Z., Zhang, H., Gong, P.T.: DualGAN: Unsupervised dual learning for image-to-image translation. In: Proc. Int. Conf. Computer Vision, pp. 2868–2876 (2017)
63. Yu, J., Lin, Z., Yang, J., Shen, X., Lu, X., Huang, T.S.: Generative image inpainting with contextual attention. In: Proc. IEEE Conf. Computer Vision and Pattern Recognition, pp. 1–15 (2018)
64. Zhou, T., Brown, M., Snavely, N., Lowe, D.G.: Unsupervised learning of depth and ego-motion from video. In: Proc. Conf. Computer Vision and Pattern Recognition, pp. 6612–6619 (2017)
65. Zhou, T., Krahenbuhl, P., Aubry, M., Huang, Q., Efros, A.A.: Learning dense correspondence via 3D-guided cycle consistency. In: Proc. IEEE Conf. Computer Vision and Pattern Recognition, pp. 117–126 (2016)
66. Zhu, J.Y., Park, T., Isola, P., Efros, A.A.: Unpaired image-to-image translation using cycle-consistent adversarial networks. In: Proc. Int. Conf. Computer Vision, pp. 2242–2251 (2017)
67. Zhuo, W., Salzmann, M., He, X., Liu, M.: Indoor scene structure analysis for single image depth estimation. In: Proc. IEEE Conf. Computer Vision and Pattern Recognition, pp. 614–622 (2015)

Part IV
Future Directions in Domain Adaptation

Chapter 9
Towards Scalable Image Classifier Learning with Noisy Labels via Domain Adaptation

Kuang-Huei Lee, Xiaodong He, Linjun Yang, and Lei Zhang

9.1 Background

One of the key factors that has been driving recent advances in large-scale image recognition is massive collections of labeled images like ImageNet [6] and MSCOCO [19]. However, it is normally expensive and time-consuming to collect large-scale manually labeled datasets. For fine-grained recognition in niche domains, data collection could be even more challenging as it sometimes requires recruiting a team of domain experts to label data [2, 22, 39, 44].

In practice, for fast development of image recognition models, it is common to use surrogates such as

- Web images with user-provided labels. For example, YFCC100M dataset [34] was collected from Flickr where users provide text labels to uploaded images. The user-provided labels are usually noisy.
- Web images crawled using image search engines with a fixed set of vocabulary as search queries, where each query is associated with an image class, e.g. WebVision dataset [17] was crawled using Google and Flickr image search with

K.-H. Lee (✉)
Microsoft AI and Research, Redmond, WA, USA

Google Brain, San Francisco, CA, USA

X. He
JD AI Research, Beijing, China
e-mail: xiaodong.he@jd.com

L. Yang
Facebook, Seattle, WA, USA

L. Zhang
Microsoft Research, Redmond, WA, USA
e-mail: leizhang@microsoft.com

© Springer Nature Switzerland AG 2020
H. Venkateswara, S. Panchanathan (eds.), *Domain Adaptation in Computer Vision with Deep Learning*, https://doi.org/10.1007/978-3-030-45529-3_9

search queries generated from the 1000 ILSVRC synsets [6]; Krause et al. [12] obtained training images and labels for several fine-grained recognition tasks using Google Image Search.

- Machine generated image labels, e.g. OpenImages dataset [14] was originally created by using Google Cloud Vision API to label a collection of annotation-free Flickr images.

These methods are easy to scale yet inevitably introduce noisy labels and domain shifts (which may also be treated as label noise.) Many studies have shown that noisy labels can affect accuracy of the induced classifiers significantly [7, 24, 27, 31], making it desirable to develop algorithms for learning in presence of noisy labels.

9.2 The Conflict Between Scalability and Effectiveness of Human Supervision

We can roughly group approaches that were previously proposed for learning classifiers with noisy labels by whether human efforts (or other kind of high-accuracy supervision signal) are involved. The simplest approach that involves human is having labeling workers review the whole dataset (e.g. images scraped from Internet) and remove mislabeled instances. A well-known example is the ImageNet dataset [6]. Some of the large-scale training data such as LSUN dataset [46] and Places dataset [47] were constructed using image classification algorithms in combination with humans to label the images semi-automatically, amplifying human effort. In the LSUN workflow, for example, the class label verification process for each class is treated as a binary classification problem—i.e., every image collected for a specific class is marked as either positive or negative. A binary classifier is learned for each image class on a small set of manually judged seed images, and run on the full unverified pool to predict whether an image is mislabeled. On the other hand, some studies for learning convolutional neural networks (CNNs) with noisy labels also rely on manual labeling to estimate label confusion and subsequently make correction [25, 43]. These methods all exhibit a disadvantage in scalability as they require human labeling effort for every image class. For classification tasks with hundreds of thousands of classes or even more (also known as extreme classification tasks) [5, 8, 41], it is nearly infeasible to have even one manual annotation per class.

In contrast, approaches that do not rely on human efforts, such as model predictions-based filtering [7] and unsupervised outliers removal [20, 30, 42], are scalable but often less effective and more heuristic due to the lack of reliable supervision. For example, unsupervised outliers removal assumes that outliers are mislabeled instances. However, in practice, outliers are often not well-defined and even not related to mislabeling (e.g. rare instances, which could even be valuable for model learning). Removing them presents a challenge [7].

Taking any of these approaches, either all the classes or none need to be manually verified, leading to a conflict between scalability and the effectiveness of human supervision.

9.3 Tackle Noisy Labels at Scale Through Domain Adaptation

Transfer learning reconciles the conflict by relaxing the requirement of human labeling efforts. In the transfer learning set-up for learning classifiers with noisy labels, manual labeling is only needed for some classes, whereas classes not covered with human supervision rely on domain adaptation.

Consider the multiclass classification setting for simplicity. Assume we have a dataset of n images, i.e., $X = \{(x_1, y_1), \ldots, (x_n, y_n)\}$, where x_i is the i-th image and $y_i \in \{1, \ldots, L\}$ is its class label, where L is the total number of classes. The class labels are noisy, which means some of the images are not related to their class labels.

Each image and its class label are marked by a **verification label**, defined as

$$l = \begin{cases} 1 & \text{if the image is correct for its class label} \\ 0 & \text{if the image is incorrect for its class label} \\ -1 & \text{if the image is not verified} \end{cases} \tag{9.1}$$

The manual labeling process to obtain valid verification labels is called **verification**. Essentially, each image is now associated with two labels: a class label and a verification label. For convenience, we refer $l = -1$ as verification labels not available, and also use the terms "label" and "class label" interchangeably.

In order to minimize human efforts, most images are left unverified (their verification labels are -1), and many classes may not have any verified data. The goal is then to develop models that learn from verification labels obtained for some image classes, and adapt models to other classes that are not covered with human supervision. This would effectively address the scalability problem of using human supervision to tackle noisy labels.

9.4 CleanNet

CleanNet (label **Clean**ing **Net**work) [16] is the first model that practically implements the transfer learning set-up for learning classifiers with noisy labels. It was influenced by an observation that many methods that tackle noisy labels explicitly or implicitly create *class prototypes* to effectively represent image classes. Methods learn from manually verified seed images [46, 47] and methods assume majority correctness like [1] belong to this category. CleanNet extends the idea of creating class prototypes from a transfer learning perspective—it learns to select representative seed images in noisy **reference image sets**[1] from human supervision, and transfers the knowledge of "how to select" to classes without human supervision signal.

CleanNet tackles noisy labels in two steps: (1) distinguishing noisy labels and (2) learning image classifiers with training images weighted based on how likely an image is mislabeled. In Sect. 9.4.1, we detail the CleanNet model architecture and learning objectives. In Sects. 9.4.2 and 9.4.3 we describe how to apply CleanNet to identifying noisy labels and learning image classifiers with noisy labels, respectively.

9.4.1 The CleanNet Model

9.4.1.1 Overview

The overall architecture of CleanNet is illustrated in Fig. 9.1. It is a joint neural embedding network that consists of two parts: a **reference set encoder** and a **query encoder**.

The reference set encoder $f_s(\cdot)$ (detailed in Sect. 9.4.1.2) takes as input a reference image set that represents a class. The reference set images are transformed into feature vectors using a CNN $f_v(\cdot)$ pretrained on noisy data. The reference set encoder learns to attend on representative image features in a noisy reference image set, and fuses them into a class-level embedding vector (class embedding). In practice, using all images in a reference set as input to the reference set encoder could be computationally expensive. Consider a reference set for class c. A pragmatic alternative is running K-means on the features of all images in the reference set to find K cluster centroids and subsequently form a smaller "reference feature set", denoted by V_c^s. The class embedding is then created from the reference feature set:

$$\phi_c^s = f_s(V_c^s) \tag{9.2}$$

The other component of CleanNet is the query encoder $f_q(\cdot)$ (detailed in Sect. 9.4.1.3). Let q denote a query image labeled as class c. Similar to reference images, the query image is also transformed into a feature vector using the same pretrained CNN:

---

[1]A reference image set is a noisy set of images associated with an image class. For example, assuming "dog" is a class label in a classification task, images scraped for "dog" from internet can be used as its reference set. For CleanNet, a reference set is created for each class as its "reference". Note that there is no human supervision from reference sets.

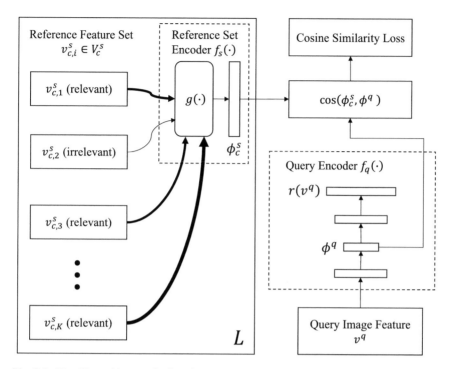

Fig. 9.1 CleanNet architecture for learning a class embedding vector ϕ_c^s and a query embedding vector ϕ_q with a similarity matching constraint. There exists one class embedding for each of the L classes. Details of the attention component $g(\cdot)$ are depicted in Fig. 9.2. The parameters of the reference set encoder including the context vector u are tied across all classes to enable knowledge transfer

$$v^q = f_v(q) \tag{9.3}$$

The query encoder $f_q(\cdot)$ then maps the query image feature v^q to a query embedding:

$$\phi^q = f_q(v^q) \tag{9.4}$$

The query embedding ϕ^q and class embedding ϕ_c^s are subject to a matching constraint, where ϕ^q is encouraged to come close to the corresponding class embedding ϕ_c^s if the query q is relevant to its class label c; otherwise, the similarity between the two embedding vectors is minimized. In other words, we decide whether a query image is mislabeled by comparing the query embedding vector with the corresponding class embedding vector. As the description implies, this matching constraint is mainly driven by supervision from verification labels (See Sects. 9.4.1.4 and 9.4.1.5.)

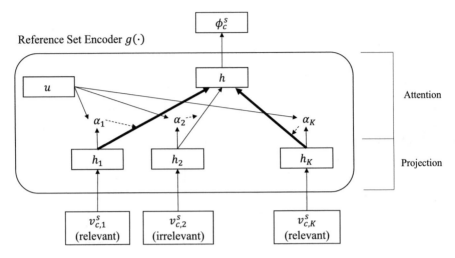

Fig. 9.2 Reference set encoder $f_s(\cdot)$ attends to reference image features deferentially and fuses the them into a class-level embedding ϕ_c^s that represents a class

9.4.1.2 Reference Set Encoder

The reference set encoder architecture is depicted in Fig. 9.2. As described previously, it maps a reference feature set V_c^s for class c to a class embedding vector ϕ_c^s. A two-layer multilayer perceptron (MLP) projects each reference feature to a hidden representation h_i. Then a dot-product attention mechanism [21] selects the hidden representations deferentially according to their importance and fuses them into a fixed-length representation h:

$$u_i = tanh(Wh_i + b) \tag{9.5}$$

$$\alpha_i = \frac{exp(u_i^T u)}{\sum_i exp(u_i^T u)} \tag{9.6}$$

$$h = \sum_i \alpha_i h_i \tag{9.7}$$

As shown in Eq. (9.7), the importance of each h_i is measured by the similarity between u_i and a context vector u. The context vector u is randomly initialized and learned during training. Driven by the matching constraint (detailed in Sects. 9.4.1.4 and 9.4.1.5), this attention mechanism learns how to focus on the most representative features for a class. Finally, a fully-connected layer maps the representation h to a class-level embedding ϕ_c^s that represents the class. The parameters of the reference set encoder including the context vector u are tied across all classes to enable knowledge transfer.

9.4.1.3   Query Encoder

The query encoder of CleanNet is a five-layer auto-encoder [10], as shown in Fig. 9.1. A typical auto-encoder usually consists of a symmetric encoder-decoder architecture. It compresses input vector into a compact representation in the middle and then seeks to reconstruct the original input, driven by a unsupervised reconstruction loss. The auto-encoder benefits the learning process by incorporating all images from all classes for training, including the majority without verification labels. Enforcing the reconstruction loss jointly with the matching constraint encourages query embeddings to preserve rich semantic information of all image classes regardless whether supervision exists or not, and subsequently enhances performance of transfer learning.

Given a query image feature vector v^q, the auto-encoder maps v^q to a hidden representation ϕ^q and seek to reconstruct v^q from ϕ^q. The reconstruction loss is defined as

$$L_r(v^q) = ||v^q - r(v^q)||^2 \tag{9.8}$$

where $r(v^q)$ is the reconstructed representation.

9.4.1.4   Supervised Matching Constraint

The matching constraint is primarily driven by supervision from verification labels. For a verified query image associated with class label c, the similarity between the corresponding class embedding ϕ_c^s and the query embedding ϕ^q is maximized if the query image is relevant to class c (verification label $l = 1$); otherwise the similarity is minimized ($l = 0$). CleanNet adopts a cosine similarity loss with margin to impose this constraint:

$$L_{cos}(\phi^q, \phi_c^s, l) = \begin{cases} 1 - \cos(\phi^q, \phi_c^s) & \text{if } l = 1 \\ \omega(max(0, \cos(\phi^q, \phi_c^s) - \rho)) & \text{if } l = 0 \\ 0 & \text{if } l = -1 \end{cases} \tag{9.9}$$

where ω is a negative sample weight for balancing positive and negative samples, ρ is the margin, and cos is the normalized cosine similarity function:

$$\cos(\phi^q, \phi_c^s) = \frac{\phi^q \cdot \phi_c^s}{||\phi^q||||\phi_c^s||} \tag{9.10}$$

Unverified images ($l = -1$) are ignored in Eq. (9.9) since this supervised objective only considers query images with verification label.

9.4.1.5 Unsupervised Matching Constraint Based on Self-Training

Handling the domain shifts in adaptation to image classes without explicit supervision relies on an unsupervised self-training objective, which introduces pseudo-
verification labels to unverified images. Specifically, an unverified image with class
label c is treated as a relevant query image if $\cos(\phi^q, \phi_c^s)$ is larger than the margin
ρ:

$$L_{cos}^{unsup}(\phi^q, \phi_c^s) = \begin{cases} 1 - \cos(\phi^q, \phi_c^s) & \text{if } l_{sudo} = 1 \\ 0 & \text{if } l_{sudo} = 0 \end{cases} \quad (9.11)$$

$$l_{sudo} = \begin{cases} 1 & \text{if } \cos(\phi^q, \phi_c^s) \geq \rho \\ 0 & otherwise \end{cases} \quad (9.12)$$

where ρ is the same margin as in Eq. (9.9). From Eqs. (9.11) and (9.12), it can be
observed that for queries that are initially treated as relevant, the model learns to
further push up the similarity between queries and reference sets; for queries that
are initially treated as irrelevant, they are ignored.

Self-training [23, 45] is an simple yet effective approach to semi-supervised
learning and unsupervised domain adaptation. As the name implies, it leverages a
model's own predictions on unlabelled data as a proxy for supervision as we have in
Eqs. (9.11) and (9.12). Typically the most confident predictions are taken as pseudo-
labels.

Self-training has shown mixed success in unsupervised domain adaptation
[26, 29, 36] and many other tasks in computer vision [28] and natural language
processing [11, 32, 38]. Lee [15] argued that the effect of training a classifier with
pseudo-labels is equivalent to entropy regularization. Although Eq. (9.11) is not in
the form of entropy, involving it in training loss still implicitly minimizes entropy
as it reduces uncertainty of model predictions, especially for target domains (classes
without human supervision signal).

9.4.1.6 The Complete Learning Objectives

To summarize the training objectives, CleanNet is learned by minimizing the following loss function combining the supervised matching constraint, the unsupervised
reconstruction loss, and the unsupervised matching constraint based on self-training:

$$L_{total} = L_{cos} + \beta L_r + t\gamma L_{cos}^{unsup} \quad (9.13)$$

$$t = \begin{cases} 1 & \text{if } l = -1 \\ 0 & \text{if } l \in \{0, 1\} \end{cases} \quad (9.14)$$

where β and γ are coefficients for the unsupervised reconstruction loss and the unsupervised matching constraint, and t indicates whether a query image has verification label. γ is usually chosen to be less than 1 to emphasize the supervised matching constraint.

9.4.2   CleanNet for Detecting Label Noise

From a relevance perspective, CleanNet can be used to rank all the images collected for a class by cosine similarity $\cos(\phi^q, \phi_c^s)$ which indicates the relevance between the query image and its class label. We can simply put a threshold on the cosine similarity to predict whether an image is relevant to its class label, namely to predict the verification label, for label noise detection:

$$\hat{l} = \begin{cases} 1 & \text{if } \cos(\phi^q, \phi_c^s) \geq \delta \\ 0 & otherwise \end{cases} \tag{9.15}$$

where δ is a threshold that can be selected through cross-validation.

Figure 9.3 compares the label noise detection results of CleanNet with other baseline methods on the Food-101N dataset [16]. Food-101N is a benchmark dataset for label noise detection and learning image classification with noisy labels. It contains 310,009 noisy images collected from Google image search, Bing image search, Yelp, and TripAdvisor using the taxonomy of Food-101 dataset [3]. There are 52,868 images with verification labels for training and 4741 images with verification labels for testing label noise detection. Each class contains about the same amount of verification labels. Image classifiers trained on Food-101N is evaluated on the original Food-101 test set.

Figure 9.3a compares the label noise detection results of CleanNet, MLPs [46], and classification filtering [35]. Specifically, the CleanNet result in Fig. 9.3a is based on using verification labels in all classes to provide human supervision. MLPs are binary classifiers trained on all verification labels for each class, serving as a baseline method which uses human supervision. Classification filtering is weakly supervised and presented as a baseline method that does not require additional human supervision[2]: With an image classifier pretrained on noisy data, an image is categorized as relevant to its class label if the class is in top-K predictions (K is selected through cross-validation), and otherwise is categorized as mislabeled. The results show that CleanNet is comparable to MLPs when all classes are provided with human supervision.

By excluding verification labels in some classes for training CleanNet, we can demonstrate the effectiveness of knowledge transfer. Figure 9.3b shows how

---

[2]As reported in [16], classification filtering is the best among the baseline methods that do not need additional human supervision, including various unsupervised outliers removal approaches.

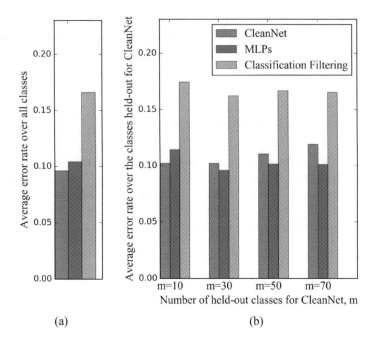

(a)   (b)

Fig. 9.3 Label noise detection results on Food-101N in terms of average error rates (%). (a) All verification labels are used for learning CleanNet and MLPs; no verification label is required for classification filtering. (b) Verification labels in $m/101$ classes are held out for learning CleanNet, whereas MLPs still use all verification labels. In (b) average error rates of all three methods are ONLY evaluated on m classes held out for CleanNet (note that MLPs can still access all verification labels)

the average error rate of label noise detection changes with verification labels in $m/101$ classes being held out for learning CleanNet, where the average error rate of CleanNet is only evaluated on the held-out classes to focus on measuring the effectiveness of knowledge transfer. CleanNet are again compared with MLPs and classification filtering in terms of average error rate evaluated on the classes held out for learning CleanNet. Note that MLPs can still access all verification labels so that we have an idea what if all the verification labels are available to a simple baseline approach. It can be observed that CleanNet is comparable to MLPs on the held-out classes, even when 70 classes are excluded.

9.4.3 CleanNet for Learning Classifiers with Noisy Labels

Now we consider applying CleanNet to learning classifiers with noisy labels. CleanNet predicts the relevance of an image to its noisy class label by comparing the query embedding of the image to the class embedding. That is, the distance between two embeddings can be used to decide how much attention we should pay to a data sample in training the image classifier.

Specifically, data samples are weighted based on the cosine similarity predicted by CleanNet:

$$w_{soft}(x, y) = \left[max(0, cos(f_q(f_v(x)), f_s(V_c^s))) \Big|_{y=c} \right] \qquad (9.16)$$

where V_c^s is the corresponding reference feature set for class c. Equation (9.16) defines a soft weighting on an image x with class label $y = c$. Similarly, we can also perform a hard weighting:

$$w_{hard}(x, y) = \begin{cases} 1 & \text{if } \left[cos(f_q(f_v(x)), f_s(V_c^s)) \Big|_{y=c} \right] \geq \delta \\ 0 & otherwise \end{cases} \qquad (9.17)$$

where δ is a threshold as in Eq. (9.15). In essence, hard weighting is equivalent to explicit label noise removal. According to the experimental results from Lee et al. [16], soft weighting is more effective, and thus we focus on using soft weighting in the following.

The classification learning objective is defined as a weighted cross-entropy loss:

$$L_{weighted} = \sum_{i=0}^{n} \left[w_{soft}(x_i, y_i, V_{y_i}^s) l(x_i, y_i) \right] \qquad (9.18)$$

where $l(x_i, y_i)$ is negative log likelihood for a training sample x_i with class label y_i:

$$l(x_i, y_i) = -log\,\hat{p}_\theta(y_i|x_i) \qquad (9.19)$$

Learning the image classifiers relies on CleanNet to assign proper weights to data samples. On the other hand, a better CNN classifier provide more discriminative image representations which are critical for CleanNet learning. To exploit this mutually beneficial relationship, CleanNet can be integrated with a CNN classifier into one framework and learned with following alternative steps:

Step 1   Train a CNN classifier on noisy data with all sample weights set to 1 till convergence.

Step 2   Fix the CNN parameters and use it to extract image features for learning CleanNet, and train CleanNet till convergence.

Step 3   Fix the CleanNet parameters and use it to weight training samples, and train the CNN classifier till convergence.

Steps 2 and 3 for learning CleanNet and the CNN classifier can be alternatively continued till both models stop improving.

Figure 9.4 presents how the image classification accuracy (using ResNet-50 [9]) changes with verification labels in $m/101$ classes being held out. Comparing to

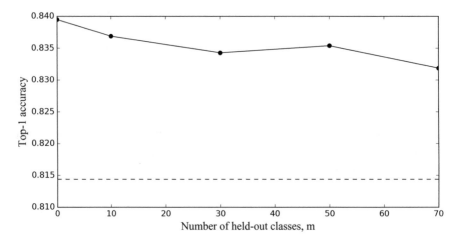

Fig. 9.4 Image classification on Food-101N in terms of top-1 accuracy (%). The solid line shows the results when verification labels in $m/101$ classes are held out for learning CleanNet. The dashed line shows the baseline classification results without using CleanNet—i.e. directly training classifiers on the noisy data

directly learning the classifier on noisy data, using CleanNet significantly improves the results even when verification labels in 70/101 classes are excluded from training.

9.5 Looking Beyond

CleanNet marks the current progress of the transfer learning paradigm for learning image classifiers with noisy labels. Successful adaptation to classes without explicit human supervision can be largely credited to the self-training strategy (and partially to the auto-encoder). That is to say, although the reference set encoder can take as input any reference set that represent a class at inference time, it does not generalize well to classes that are not seen at training time and not learned through the unsupervised objectives. Thus, CleanNet is constrained by a predefined set of classes once it is trained. How to remove this limitation remains an open issue. The ability of adapting to classes unseen at training time could be beneficial as it (1) reduces overheads of developing new classification models, and (2) enables scenarios where novel image classes cannot be predicted and included at training time, e.g. filtering noisy images in image search results (an image search query can be viewed as defining an image class but search queries are unbounded.)

On the other hand, CleanNet has been developed as a neural joint embedding network, mapping reference sets and queries to representations in joint semantic embedding space. Learning CleanNet was formulated as a metric learning problem. As described in Sect. 9.4, distance in the joint semantic embedding space has been utilized for (1) identifying noisy labels through thresholding and (2) weighting classification training samples. In contrast to metric learning, using discriminative models with probabilistic prediction (e.g. [37]) exhibits an advantage in having probabilistic interpretation. Identifying noisy labels or weighting training samples based on probabilistic prediction could be more interpretable, and may be combined with statistical prior and more complicated probabilistic formation.

At last, the transfer learning paradigm discussed in this chapter is not limited to multiclass image classification tasks. Generalizing to multilabel classification is straight-forward as it does not require an image to be tied to only one label. The set-up and methods that we discussed in this chapter are also applicable to classification tasks as long as proper representations are provided.

9.6 Related Works

In this last part of the chapter we briefly discuss related works regarding to detecting mislabeled instances in data and learning classifiers with noisy labels. Some of them have been mentioned in previous sections.

9.6.1 Detecting Mislabeled Instances in Data

As described in Sect. 9.4, CleanNet belongs to the category of approaches that tackles noisy labels for classifier learning in two steps: first distinguishing mis-labeled instances, and then down-weighting or removing mislabeled instances for training classification models. One of the popular approaches in this category is unsupervised outliers detection and removal (e.g. One-Class SVM [30], UOCL [20], and DRAE [42]) by assuming that outliers are mislabeled instances. However, as discussed earlier in Sect. 9.1, outliers are often not well defined. The fundamental assumption often does not hold, and therefore removing them presents a challenge [7]. Another approach that does not require human efforts is weakly supervised label noise reduction [7]. For example, Thongkam et al. [35] proposed a classification filtering method that learns an SVM from noisy data and removes instances misclassified by the SVM. Weakly supervised methods are often heuristic and less effective without reliable supervision. On the other hand, learning to detect label noise from human supervision has been widely studied for dataset constructions. For instance, Yu et al. [46] proposed training MLPs on manually labeled seed images and using them to remove mislabeled images. Similarly, the Places dataset [47] was created using an AlexNet [13] trained on manually verified seed images. However, methods using human supervision exhibit a disadvantage in scalability as they require human supervision for all the image classes. It leads the motivation behind the transfer learning set-up that we discuss in this chapter.

9.6.2   Learning Image Classifiers with Label Noise End-to-End

Some methods were developed for learning neural network with label noise end-to-end [1, 4, 18, 25, 27, 31, 33, 40, 43, 48], including those relying on human supervision for all classes and those do not require human labeling efforts. These methods are again facing the same conflict between effectiveness and scalability.

A group of works, e.g. the methods by Xiao et al. [43] and Patrini et al. [25], proposed handling noisy labels with confusion matrix estimated on manually labeled seed images. In such labeling process, seed images are augmented with manually assigned class labels in addition to noisy class labels. Different from **verification** that we introduced in Sect. 9.3, this is another type of human labeling efforts for tackling noisy labels that can be seen in literature, and we call it **correction**. However, it presents two major limitations comparing to verification. First and most importantly, for a classification task involving a large set of class labels, manually selecting the correct class label for an image is much more challenging than simply verifying whether the noisy class label is relevant. Extending the labeling efforts for correction to a large set of classes is practically intractable. Moreover, it is not clear how to extend label correction to multilabel classification as it requires an image to be tied with an ground-truth class label.

References

1. Azadi, S., Feng, J., Jegelka, S., Darrell, T.: Auxiliary image regularization for deep CNNs with noisy labels. In: International Conference on Learning Representations (2016)
2. Bejnordi, B.E., Veta, M., Van Diest, P.J., Van Ginneken, B., Karssemeijer, N., Litjens, G., Van Der Laak, J.A., Hermsen, M., Manson, Q.F., Balkenhol, M., et al.: Diagnostic assessment of deep learning algorithms for detection of lymph node metastases in women with breast cancer. Jama 318(22), 2199–2210 (2017)
3. Bossard, L., Guillaumin, M., Van Gool, L.: Food-101–mining discriminative components with random forests. In: European Conference on Computer Vision (2014)
4. Chen, X., Gupta, A.: Webly supervised learning of convolutional networks. In: International Conference on Computer Vision (2015)
5. Dean, T., Ruzon, M.A., Segal, M., Shlens, J., Vijayanarasimhan, S., Yagnik, J.: Fast, accurate detection of 100,000 object classes on a single machine. In: Proceedings of the IEEE Conference on Computer Vision and Pattern Recognition (2013)
6. Deng, J., Dong, W., Socher, R., Li, L.J., Li, K., Fei-Fei, L.: ImageNet: A large-scale hierarchical image database. In: 2009 IEEE Conference on Computer Vision and Pattern Recognition (2009)
7. Frénay, B., Verleysen, M.: Classification in the presence of label noise: a survey. IEEE Trans. Neur. Net. Lear. Sys. 25(5), 845–869 (2014)
8. Frome, A., Corrado, G.S., Shlens, J., Bengio, S., Dean, J., Mikolov, T., et al.: Devise: A deep visual-semantic embedding model. In: Conference on Neural Information Processing Systems (2013)
9. He, K., Zhang, X., Ren, S., Sun, J.: Deep residual learning for image recognition. In: Proceedings of the IEEE Conference on Computer Vision and Pattern Recognition (2016)

10. Hinton, G.E., Salakhutdinov, R.R.: Reducing the dimensionality of data with neural networks. Science **313**(5786), 504–507 (2006)
11. Huang, Z., Harper, M.: Self-training PCFG grammars with latent annotations across languages. In: Proceedings of the 2009 Conference on Empirical Methods in Natural Language Processing (2009)
12. Krause, J., Sapp, B., Howard, A., Zhou, H., Toshev, A., Duerig, T., Philbin, J., Fei-Fei, L.: The unreasonable effectiveness of noisy data for fine-grained recognition. In: European Conference on Computer Vision (2016)
13. Krizhevsky, A., Sutskever, I., Hinton, G.E.: ImageNet classification with deep convolutional neural networks. In: Advances in Neural Information Processing Systems (2012)
14. Kuznetsova, A., Rom, H., Alldrin, N., Uijlings, J., Krasin, I., Pont-Tuset, J., Kamali, S., Popov, S., Malloci, M., Duerig, T., Ferrari, V.: The open images dataset v4: Unified image classification, object detection, and visual relationship detection at scale (2018). arXiv:1811.00982
15. Lee, D.H.: Pseudo-label: The simple and efficient semi-supervised learning method for deep neural networks. In: Workshop on Challenges in Representation Learning, ICML (2013)
16. Lee, K.H., He, X., Zhang, L., Yang, L.: CleanNet: Transfer learning for scalable image classifier training with label noise. In: Proceedings of the IEEE Conference on Computer Vision and Pattern Recognition (2018)
17. Li, W., Wang, L., Li, W., Agustsson, E., Van Gool, L.: Webvision database: Visual learning and understanding from web data (2017). Preprint. arXiv:1708.02862
18. Li, Y., Yang, J., Song, Y., Cao, L., Luo, J., Li, J.: Learning from noisy labels with distillation. In: Proceedings of the IEEE International Conference on Computer Vision (2017)
19. Lin, T.Y., Maire, M., Belongie, S., Hays, J., Perona, P., Ramanan, D., Dollár, P., Zitnick, C.L.: Microsoft COCO: Common objects in context. In: European Conference on Computer Vision (2014)
20. Liu, W., Hua, G., Smith, J.R.: Unsupervised one-class learning for automatic outlier removal. In: Proceedings of the IEEE Conference on Computer Vision and Pattern Recognition (2014)
21. Luong, M.T., Pham, H., Manning, C.D.: Effective approaches to attention-based neural machine translation. In: Conference on Empirical Methods in Natural Language Processing (2015)
22. Maji, S., Rahtu, E., Kannala, J., Blaschko, M., Vedaldi, A.: Fine-grained visual classification of aircraft (2013). Preprint. arXiv:1306.5151
23. McClosky, D., Charniak, E., Johnson, M.: Effective self-training for parsing. In: Proceedings of the Main Conference on Human Language Technology Conference of the North American Chapter of the Association of Computational Linguistics (2006)
24. Nettleton, D.F., Orriols-Puig, A., Fornells, A.: A study of the effect of different types of noise on the precision of supervised learning techniques. Artif. Intell. Rev. **33**(4), 275–306 (2010)
25. Patrini, G., Rozza, A., Krishna Menon, A., Nock, R., Qu, L.: Making deep neural networks robust to label noise: A loss correction approach. In: Proceedings of the IEEE Conference on Computer Vision and Pattern Recognition (2017)
26. Reichart, R., Rappoport, A.: Self-training for enhancement and domain adaptation of statistical parsers trained on small datasets. In: Proceedings of the 45th Annual Meeting of the Association of Computational Linguistics (2007)
27. Rolnick, D., Veit, A., Belongie, S., Shavit, N.: Deep learning is robust to massive label noise (2017). Preprint. arXiv:1705.10694
28. Rosenberg, C., Hebert, M., Schneiderman, H.: Semi-supervised self-training of object detection models. In: IEEE Workshop on Applications of Computer Vision (2005)
29. Saito, K., Ushiku, Y., Harada, T.: Asymmetric tri-training for unsupervised domain adaptation. In: Proceedings of the 34th International Conference on Machine Learning (2017)
30. Schölkopf, B., Platt, J.C., Shawe-Taylor, J., Smola, A.J., Williamson, R.C.: Estimating the support of a high-dimensional distribution. Neural Comput. **13**(7), 1443–1471 (2001)
31. Sukhbaatar, S., Bruna, J., Paluri, M., Bourdev, L., Fergus, R.: Training convolutional networks with noisy labels (2014). Preprint. arXiv:1406.2080

32. Suzuki, J., Isozaki, H.: Semi-supervised sequential labeling and segmentation using giga-word scale unlabeled data. In: Proceedings of the Main Conference on Human Language Technology Conference of the North American Chapter of the Association of Computational Linguistics (2008)
33. Tanaka, D., Ikami, D., Yamasaki, T., Aizawa, K.: Joint optimization framework for learning with noisy labels. In: Proceedings of the IEEE Conference on Computer Vision and Pattern Recognition (2018)
34. Thomee, B., Shamma, D.A., Friedland, G., Elizalde, B., Ni, K., Poland, D., Borth, D., Li, L.J.: YFCC100M: The new data in multimedia research. Commun. ACM **59**(2), 64–73 (2016)
35. Thongkam, J., Xu, G., Zhang, Y., Huang, F.: Support vector machine for outlier detection in breast cancer survivability prediction. In: Asia-Pacific Web Conference, pp. 99–109. Springer, Berlin (2008)
36. Tsai, Y.H.H., Huang, L.K., Salakhutdinov, R.: Learning robust visual-semantic embeddings. In: 2017 IEEE International Conference on Computer Vision (2017)
37. Tzeng, E., Hoffman, J., Saenko, K., Darrell, T.: Adversarial discriminative domain adaptation. In: Proceedings of the IEEE Conference on Computer Vision and Pattern Recognition (2017)
38. Van Asch, V., Daelemans, W.: Predicting the effectiveness of self-training: Application to sentiment classification (2016). Preprint. arXiv:1601.03288
39. Van Horn, G., Branson, S., Farrell, R., Haber, S., Barry, J., Ipeirotis, P., Perona, P., Belongie, S.: Building a bird recognition app and large scale dataset with citizen scientists: The fine print in fine-grained dataset collection. In: Proceedings of the IEEE Conference on Computer Vision and Pattern Recognition (2015)
40. Veit, A., Alldrin, N., Chechik, G., Krasin, I., Gupta, A., Belongie, S.: Learning from noisy large-scale datasets with minimal supervision. In: Proceedings of the IEEE Conference on Computer Vision and Pattern Recognition (2017)
41. Weston, J., Bengio, S., Usunier, N.: Wsabie: Scaling up to large vocabulary image annotation. In: Twenty-Second International Joint Conference on Artificial Intelligence (2011)
42. Xia, Y., Cao, X., Wen, F., Hua, G., Sun, J.: Learning discriminative reconstructions for unsupervised outlier removal. In: Proceedings of the IEEE International Conference on Computer Vision (2015)
43. Xiao, T., Xia, T., Yang, Y., Huang, C., Wang, X.: Learning from massive noisy labeled data for image classification. In: Proceedings of the IEEE Conference on Computer Vision and Pattern Recognition (2015)
44. Yan, K., Wang, X., Lu, L., Summers, R.M.: DeepLesion: automated mining of large-scale lesion annotations and universal lesion detection with deep learning. J. Med. Imag. **5**(3), 1–11–11 (2018). https://doi.org/10.1117/1.JMI.5.3.036501
45. Yarowsky, D.: Unsupervised word sense disambiguation rivaling supervised methods. In: 33rd Annual Meeting of the Association for Computational Linguistics (1995)
46. Yu, F., Seff, A., Zhang, Y., Song, S., Funkhouser, T., Xiao, J.: LSUN: Construction of a large-scale image dataset using deep learning with humans in the loop (2015). Preprint. arXiv:1506.03365
47. Zhou, B., Lapedriza, A., Khosla, A., Oliva, A., Torralba, A.: Places: A 10 million image database for scene recognition. IEEE Trans. Pattern Anal. Mach. Intell. **40**, 1452–1464 (2017)
48. Zhuang, B., Liu, L., Li, Y., Shen, C., Reid, I.: Attend in groups: a weakly-supervised deep learning framework for learning from web data. In: Proceedings of the IEEE Conference on Computer Vision and Pattern Recognition (2017)

Chapter 10
Adversarial Learning Approach for Open Set Domain Adaptation

Kuniaki Saito, Shohei Yamamoto, Yoshitaka Ushiku, and Tatsuya Harada

10.1 Introduction

Deep neural networks have boosted performance on many image recognition tasks [19]. One of the challenges is that basically, they cannot recognize samples as unknown if the class of them is absent during training. We define such a class as an "unknown class" and the categories shown during training is referred to as the "known class." If these samples can be recognized as unknown, we can arrange noisy datasets and pick up the samples of interest from them. Moreover, if some systems working in the open-world can recognize unknown objects and ask humans to annotate them, these systems can easily improve their recognition ability. For these reasons, the open set recognition is very important task.

In domain adaptation, we try to train a model from a label-rich domain (source domain) and apply it to a label-poor domain (target domain). Examples in different domains have different characteristics which can harm the performance of a model. Many methods of domain adaptation are proposed with the assumption that examples in the target domain should belong to the one of the categories in the source in the upper left of Fig. 10.1. However, this assumption does not always stand. Let's take a look at the problem of an unsupervised domain adaptation, where we have only unlabeled target examples. We are not able to make sure that the target samples necessarily belong to the classes in the source because they are not annotated. The target domain can include samples of classes that are not shown in the source domain (open set domain adaptation), or the target domain may not have

K. Saito (✉) · S. Yamamoto · Y. Ushiku
The University of Tokyo, Tokyo, Japan
e-mail: k-saito@mi.t.u-tokyo.ac.jp; yamamoto@mi.t.u-tokyo.ac.jp; ushiku@mi.t.u-tokyo.ac.jp

T. Harada
The University of Tokyo and RIKEN, Tokyo, Japan
e-mail: harada@mi.t.u-tokyo.ac.jp

© Springer Nature Switzerland AG 2020
H. Venkateswara, S. Panchanathan (eds.), *Domain Adaptation in Computer Vision with Deep Learning*, https://doi.org/10.1007/978-3-030-45529-3_10

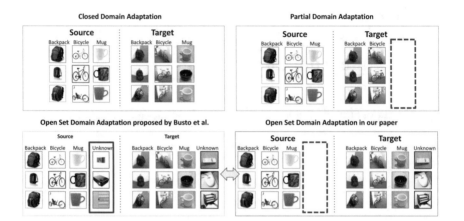

Fig. 10.1 A comparison of different domain adaptation settings. **Upper Left**: Existing closed domain adaptation. Source and target domain have the same categories. **Upper Right**: Partial domain adaptation. Source domain has examples of categories which are absent in the target domain. **Bottom Left**: Existing setting of open set domain adaptation [4]. Access to the unknown source samples is allowed. Please note that the class of unknown source does not overlap with that of unknown target. **Bottom Right**: Problem setting we will introduce in this chapter. Accessibility to the unknown samples in the source domain is not assumed

classes present in the source domain (partial domain adaptation). Algorithms that work well in such situations are very practical. To deal with this problem, the task, open set domain adaptation, was recently introduced [4] where the target domain have samples that do not belong to the class in the source domain as shown in the bottom left of Fig. 10.1. In addition, the task called a partial domain adaptation was proposed [5, 41] where the source domain has the categories which are absent in the target domain as shown in the upper right of Fig. 10.1. In this chapter, we mainly focus on the problem of open set domain adaptation. This chapter is partially based on earlier work from [31].

The solution to the open set domain adaptation needs to recognize unknown target samples as "unknown" and to classify known target samples into correct known categories. They [4] utilized unknown source samples to gain knowledge about unknown examples. However, having unknown source samples can be costly because it is necessary to collect various and many unknown source samples to obtain the concept of "unknown." Then, in this chapter, we introduce a more challenging open set domain adaptation (OSDA) that does not need any unknown source samples to train a model. That is, we have access to known source examples and unlabeled target examples for open set domain adaptation as shown in the bottom right of Fig. 10.1.

There are mainly two problems to solve this challenging problem. First, it is difficult to delineate a boundary between known and unknown classes because we do not have any labeled unknown examples. The second is the domain's difference. To correctly classify known target into known categories, we need to

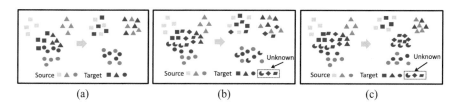

Fig. 10.2 (**a**): Closed set domain adaptation with distribution matching method. (**b**): Open set domain adaptation with distribution matching method. Unknown samples are aligned with known source samples. (**c**): Open set domain adaptation with our proposed method. OSBP enables to learn features that can reject unknown target samples

perform distribution matching between source and target. However, we should keep unknown target samples far away from known source examples. The previous distribution matching methods are proposed to match the distribution of the target with that of the source. However, this method cannot be utilized for our problem. In OSDA, it is necessary to reject unknown target samples instead of aligning them with the source.

As one example of a method to solve the problem, we introduce an approach of adversarial learning that encourages feature extractor to classify target samples into known or unknown. A comparison with previous methods is presented in Fig. 10.2. Unlike the existing feature alignment methods, this method enables the rejection of unknown target samples with high accuracy as well as the alignment of known target samples with known source samples. There are players in this method, i.e., the feature extractor and the classifier. The feature extractor outputs features from inputs images, and the classifier takes them and outputs $K+1$ dimension probability vectors, where K is the number of known categories. The $K+1$ th dimension indicates the probability for the unknown category. The classifier makes a decision boundary between source and target samples while the feature extractor is trained to put target samples far from the boundary. Specifically, the classifier is trained to output probability t for unknown class, where $0 < t < 1$. A decision boundary for unknown samples is built by weakly training the classifier to recognize all target samples as unknown. The feature extractor is trained to deceive the classifier in an adversarial manner. The feature extractor has two options here, i.e., to increase or to decrease the probability t. It means that two options are given to the feature generator: aligning them with source examples or recognizing them as unknown.

The overview of this chapter is as follows.

1. We explain the overview of the recent research on open set domain adaptation and partial domain adaptation.
2. We present the open set domain adaptation where unknown source samples are not provided.
3. We introduce an adversarial learning method for the problem. The method enables training of the feature generator to learn representations which can separate unknown target samples from known ones. This method is called OSBP
4. We show some evaluation on OSBP, which is done for digits and objects datasets.

10.2 Related Work

In this section, we briefly introduce methods for closed domain adaptation, partial domain adaptation, and open set recognition.

10.2.1 Closed Domain Adaptation

Domain adaptation for image recognition has attracted attention to transfer the knowledge between different domains and reducing the cost to give annotations to a large number of images. Benchmark datasets are provided [28], and many methods for unsupervised domain adaptation and semi-supervised domain adaptation have been proposed [8, 10–12, 22, 23, 29, 34]. As discussed in the previous sections, unsupervised and semi-supervised domain adaptation are proposed for the situation where different domains share the categories in the datasets, which may not be practical in some cases.

One of the representative methods for this tasks are distribution matching based methods [3, 8, 22, 30, 37]. Each domain has their own characteristics of their features, which can harm the performance of the model trained on each domain. Therefore, they aim to obtain domain-invariantly features by matching the distributions of features. This technique is widely used for various visual domain adaptation tasks [6, 8, 15, 32]. The representative of the methods is similar to Generative Adversarial Networks (GAN) [13]. They train a discriminator to judge whether input images are fake or real whereas the image generator is trained to deceive the discriminator. In domain adaptation, similar to GAN, the domain classifier is trained to recognize whether the features of the middle layers are from a target or a source domain whereas the feature extractor is trained to deceive it. Many variants and extensions to the generative models have been proposed [2, 3, 21, 36, 38]. Maximum Mean Discrepancy (MMD) [14] is also a popular way to measure the divergence between domains. The distance is utilized to train domain-invariantly effective models, and its variants are proposed [22–24, 39]. The methods do not perform well on an open set domain adaptation. Every target samples including unknown classes should be aligned with source if they are applied. This makes it hard to recognize unknown target samples as unknown.

By contrast, the method we introduce in this chapter enables to recognize unknown target samples as unknown, although any labeled target unknown samples are not provided during training. We will show comparison with distribution matching methods shown above. This method utilizes the technique of distribution matching to achieve open set recognition. However, this method allows the feature extractor to reject some target samples as unknown.

10.2.2   Partial Domain Adaptation

Cao et al. [5] and Zhang et al. [41] proposed a task called partial domain adaptation. Different from conventional closed domain adaptation, the target domain does not have categories which are present in the source domain. As we do not know the label of the target examples, this setting is realistic too. The overview of the setting is shown in Fig. 10.1. The difficulty in this setting is that if we try to match distributions completely, the target examples can be aligned with source categories absent in the target. Therefore, both work proposed to utilize domain classifier based adversarial learning with importance weighting. Their idea is that source examples similar to the target examples should belong to the shared categories. Focusing only on such examples should achieve better feature alignment in the partial domain adaptation setting. The difficulty of partial domain adaptation is how to find source categories which are absent in the target whereas the difficulty of OSDA is how to find target examples which should not be aligned with source examples. Although we do not present a method for partial domain adaptation in this chapter, this setting is realistic and important for many applications. In addition, the method used in partial domain adaptation can be useful for OSDA and vice versa.

10.2.3   Open Set Recognition

Many methods has been proposed to detect outliers while correctly classifying inliers during testing. Multi-class open set SVM is proposed by Jain et al. [17]. They reject unknown samples by training SVMs that can give probabilistic decision scores. They aim to reject unknown samples with a threshold probability value. In addition, a method of harnessing deep neural networks for open set recognition was proposed [1]. OpenMax layer was proposed, which accounts for the probability of an input being from an unknown class. Furthermore, to give supervision of the unknown samples in an unsupervised way, generating unknown examples was proposed [9]. The method used GAN to generate unknown samples just by seeing known examples and utilized it to train a model, then applied OpenMax layer. These methods defined a threshold value to reject unknown samples to recognize unknown samples as unknown during testing,

In this chapter, we introduce a method that tries to solve the open set recognition problem in the setting of the domain adaptation. The distribution of the known samples in the target domain is different from that of the samples in the source domain, which is different from general open set recognition and makes the task more difficult.

10.3 Method

First, we provide an overview, then explain the training procedure. The overview of the model is illustrated in Fig. 10.3.

10.3.1 Problem Setting and Key Idea

It is assumed that a source image x_s and the label y_s from a source dataset $\{X_s, Y_s\}$ are available, as well as an unlabeled target image x_t drawn from an unlabeled target dataset X_t. The source images come only from known classes whereas target ones can be from unknown class. In this method, we train a feature extractor G that takes inputs x_s or x_t, and a network C that takes features from G and categorizes them into $K + 1$ classes, where the K is the number of known categories. To summarize, C outputs a $K + 1$-dimensional vector of logits $\{l_1, l_2, l_3 \ldots l_{K+1}\}$ per one sample. We convert the logits to class probabilities using a softmax layer. Namely, the probability that x is classified into class j is shown as $p(y = j|x) = \frac{\exp(l_j)}{\sum_{k=1}^{K+1} \exp(l_k)}$. 1 $\sim K$ dimensions is the probability of the known categories while $K + 1$ dimension means the probability of the unknown class. We utilize the notation $p(y|x)$ to show the $K + 1$-dimensional probabilistic output for an input x.

Correctly categorizing known target samples into corresponding known class and recognizing unknown target samples as unknown is the goal of this method. It is necessary to make a decision boundary for the unknown class, although any information about the class is not given. Then, this method utilizes a pseudo decision boundary for unknown class, which is done by weakly training a classifier to categorize target samples as unknown. Then, feature extractor attempts to deceive

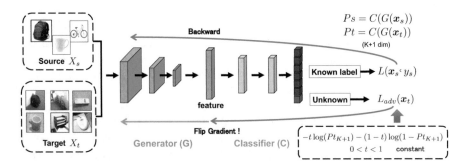

Fig. 10.3 Illustration of the method. The classifier network gives $K + 1$ dimensional probabilistic output. The network is trained to correctly classify source samples whereas, for target samples, it is trained to output t as the probability of the unknown class. On the other hand the feature extractor is trained to deceive it. This methods utilize the gradient reversal layer to achieve the adversarial training

the classifier. Importantly, feature extract needs to make unknown target samples far from known target ones. If a classifier is trained to output $p(y = K + 1|x_t) = 1.0$ and the extractor tries to deceive it, then the objective of the extractor will be to fully match the target distribution with the source one. Therefore, the extractor will only decrease the probability of unknown class. This kind of method is applied to train Generative Adversarial Networks for semi-supervised learning [33] and is expected to be useful for unsupervised domain adaptation. However, it cannot be directly applied to separate unknown samples from known ones.

Then, this method tries to train the classifier to output $p(y = K + 1|x_t) = t$, where $0 < t < 1$ and train the extractor to deceive the classifier. The objective of the feature extractor is to maximize the loss of the classifier. To increase the loss, the feature extractor can choose to increase the probability of an unknown class, which means that the example is classified as unknown. For instance, consider when t is a very small value, it should be easier for feature extractor to increase the probability of an unknown class than to decrease it to increase the loss of the classifier. Similarly, the feature extractor can choose to decrease it to make $p(y = K + 1|x_t)$ lower than t, which means that the sample is aligned with source. In summary, the feature extractor can choose whether a target sample should be matched with the source or should be rejected as unknown. In all experiments, the value of t is set as 0.5. If t is larger than 0.5, the sample is necessarily recognized as unknown class. Thus, it is assumed that this value ($t = 0.5$) can be a good decision boundary between known and unknown. In experiments, analysis of the behavior is shown when this value is varied.

10.3.2 Training Procedure

First, both the classifier and the generator should be trained to categorize source samples correctly. A standard cross-entropy loss is used for this purpose.

$$L_s(x_s, y_s) = -\log(p(y = y_s|x_s)) \tag{10.1}$$

$$p(y = y_s|x_s) = (C \circ G(x_s))_{y_s} \tag{10.2}$$

To make classifier built a boundary for an unknown class, binary cross entropy loss is applied.

$$L_{adv}(x_t) = -t \log(p(y = K + 1|x_t)) - (1 - t) \log(1 - p(y = K + 1|x_t)), \tag{10.3}$$

where t is set as 0.5. The overall objective is,

$$\min_C L_s(x_s, y_s) + L_{adv}(x_t) \tag{10.4}$$

$$\min_G L_s(x_s, y_s) - L_{adv}(x_t) \tag{10.5}$$

Algorithm 2 Mini-batch training of the method

---

for the number of training iterations **do**
  - Sample minibatch of m source samples $\{\{\boldsymbol{x}_s, y_s\}^{(1)}, \ldots, \{\boldsymbol{x}_s, y_s\}^{(m)}\}$ from $\{X_s, Y_s\}$.
  - Sample minibatch of m target samples $\{\boldsymbol{x}_t^{(1)}, \ldots, \boldsymbol{x}_t^{(m)}\}$ from X_t.
  Calculate $L_s(\boldsymbol{x}_s, y_s)$ by cross-entropy loss and $L_{adv}(\boldsymbol{x}_t)$ following Eq. 10.3.
  Update the parameter of G and C following Eq. 10.4, Eq. 10.5. Gradient reversal layer is used here.
end for

The classifier tries to make the value of $p(y = K + 1|\boldsymbol{x}_t)$ equal to t whereas the feature extractor tries to increase the value of $L_{adv}(\boldsymbol{x}_t)$. It means that the feature extractor attempts to make the value of $p(y = K + 1|\boldsymbol{x}_t)$ different from t. To calculate the gradient for $L_{adv}(\boldsymbol{x}_t)$ in an efficient way, a gradient reversal layer [8] is utilized. The layer flips of the sign of the gradient during the backward process for adversarial loss. With this layer, we can update the parameters of the classifier and generator through one back-propagation. The algorithm is summarized in Algorithm 2.

10.4 Experiments

We show the results of experiments on Office [28], VisDA [27] and digits datasets.

10.4.1 Implementation Detail

We trained the classifier and feature extractor using AlexNet [19] and VGGNet [35] pretrained on ImageNet [7]. In the experiments on both Office and VisDA dataset, the parameters of the networks are not updated. Fully-connected layers with 100 hidden units are constructed after the FC8 layers. Batch Normalization [16] and Leakly-ReLU layer were utilized. Momentum SGD with a learning rate 1.0×10^{-3} is used with the momentum equal to 0.9.

Three baselines were implemented. The first is an open set SVM (OSVM) [17]. OSVM uses a threshold to recognize samples as unknown. If the predicted probability is lower than the threshold for any class, the example should be unknown. CNN is firstly trained using only source samples, then, use it as a feature extractor. Features are extracted to train OSVM. OSVM does not utilize any unknown samples during training. Therefore, we trained OSVM using only source samples and tested them on the target samples. The second baseline is a combination with neural networks trained with Maximum Mean Discrepancy(MMD) [22] and OSVM. MMD is used to measure the distance between different domains in unsupervised domain adaptation. To adapt to an open set recognition, we trained the networks with MMD and trained

OSVM with the features extracted from the networks. A comparison with this baseline will shown how the introduced method differs from existing distribution matching methods. The final baseline is a combination with a domain classifier based method, BP [8] and OSVM. BP is a popular a distribution matching method for domain adaptation. Similar to MMD, we first trained a network with BP and extracted features to train OSVM. We utilized the same network architecture to train these baseline models. The experiments were repeated 3 times for each method, and the average score was reported. For simplicity, we call the method we introduced so far OSBP.

10.4.2   Experiments on Office

10.4.2.1   11 Class Classification

Firstly, we show the results using Office dataset following the setting propose by Busto and Gall [4]. The dataset has of 31 classes, and 10 of them were selected as shared known classes. The classes are also common in the Caltech dataset [11]. In alphabetical order, 21–31 classes are selected as unknown samples in the target. In this experiment, the model has to correctly classify samples in the target domain into 10 shared classes or unknown class. Therefore, 11 class classification was performed. Accuracy averaged over all classes is denoted as OS in tables. We also show the accuracy measured on the known classes of the target domain (OS*). Following [4], we show the accuracy averaged over the classes in the OS and OS*.

We also compared this method with a method proposed by [4]. Their method is proposed for a situation where unknown samples in the source domain are available. However, they also adapted their method using OSVM when unknown source samples were absent. To better understand the performance of methods, we also present the results which utilized the unknown source samples during training for some baselines. The numbers are cited from [4].

The results are presented in Table 10.1. Compared with the baseline methods, OSBP exhibits better performance in almost all scenarios. The accuracy of OS is better than that of OS* in many cases for OSVM, which indicates that many known target samples are regarded as unknown. This is because OSVM is designed to detect outliers and is likely to classify target samples as unknown because their characteristics are different from source. By comparing the performance of OSVM and MMD+OSVM, we can see that using MMD does not always boost the performance. The existence of unknown target samples does harm to the correct feature alignment. The features are visualized using t-SNE [25] in Fig. 10.4. Because the networks are trained with the source samples, the features are made discriminative for known classes. We can observe that the method separates unknown target samples from known ones while MMD and BP do not.

Table 10.1 Accuracy (%) of each method in 10 shared class situation

	A-D		A-W		D-A		D-W		W-A		W-D		AVG	
	OS	OS*	OS	OS*	OS	OS*	OS	OS*	OS	OS*	OS	OS*	OS	OS*
Method w/ unknown classes in source domain (AlexNet)														
BP [8]	78.3	77.3	75.9	73.8	57.6	54.1	89.8	88.9	64.0	61.8	98.7	98.0	77.4	75.7
ATI-λ [4]	79.8	79.2	77.6	76.5	71.3	70.0	93.5	93.2	76.7	76.5	98.3	99.2	82.9	82.4
Method w/o unknown classes in source domain (AlexNet)														
OSVM	59.6	59.1	57.1	55.0	14.3	5.9	44.1	39.3	13.0	4.5	62.5	59.2	40.6	37.1
MMD + OSVM	47.8	44.3	41.5	36.2	9.9	0.9	34.4	28.4	11.5	2.7	62.0	58.5	34.5	28.5
BP + OSVM	40.8	35.6	31.0	24.3	10.4	1.5	33.6	27.3	11.5	2.7	49.7	44.8	29.5	22.7
ATI-λ[4] + OSVM	72.0	-	65.3	-	**66.4**	-	82.2	-	71.6	-	92.7	-	75.0	-
Ours	**76.6**	**76.4**	**70.1**	**69.1**	62.5	**62.3**	**94.4**	**94.6**	**82.3**	**82.2**	**96.8**	**96.9**	**80.4**	**80.2**
Method w/o unknown classes in source domain (VGGNet)														
OSVM	82.1	83.9	75.9	75.8	38.0	33.1	57.8	54.4	54.5	50.7	83.6	83.3	65.3	63.5
MMD + OSVM	84.4	**85.8**	75.6	75.7	41.3	35.9	61.9	58.7	50.1	45.6	84.3	83.4	66.3	64.2
BP + OSVM	83.1	84.7	76.3	76.1	41.6	36.5	61.1	57.7	53.7	49.9	82.9	82.0	66.4	64.5
Ours	**85.8**	**85.8**	**76.9**	**76.6**	**89.4**	**91.5**	**96.0**	**96.6**	**83.4**	**83.1**	**97.1**	**97.3**	**88.0**	**88.5**

A, D and W correspond to Amazon, DSLR and Webcam respectively

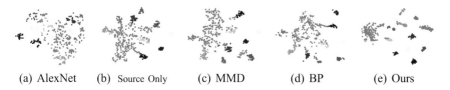

(a) AlexNet   (b) Source Only   (c) MMD   (d) BP   (e) Ours

Fig. 10.4 Visualization of target features extracted by each method. Green points are unknown target samples and different colors indicate different classes. (**a**): Features obtained by pre-trained AlexNet. (**b**): Features extracted by a model trained with no adaptation. (**c**): Features obtained by a model trained with MMD. (**d**): Features obtained by a model trained with BP. (**e**): Features obtained by OSBP. This method tries to separate unknown target samples from known target ones

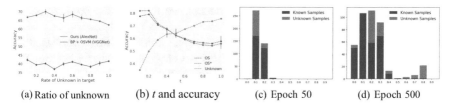

(a) Ratio of unknown   (b) t and accuracy   (c) Epoch 50   (d) Epoch 500

Fig. 10.5 (**a**): The behavior of the method when we changed the ratio of unknown samples. As the number of unknown target samples increases, the accuracy decreases. (**b**): The change of accuracy with varying the value t in the same adaptation setting. The accuracy of unknown target samples is shown as green line. As t increases, target samples are likely classified as "unknown". However, both the accuracy OS and OS* decrease. (**c**) and (**d**): Frequency diagram of the probability of unknown target examples in adaptation from Webcam to DSLR

Number of Unknown Samples and Accuracy We show the accuracy when the number of target samples varies in the adaptation from DSLR to Amazon. Unknown target samples are randomly chosen from Amazon and the ratio of the unknown samples was varied. The accuracy of OS is presented in Fig. 10.5a. When the ratio changes, OSBP seems to perform well. **Value of t** We also observe that the behavior of the model changes when the value of, t in Eq. 10.3 is varied. As we mentioned in the method section, if t is made equal to 1, the goal of the feature extractor is to fully match the whole distribution of the target features with that of the source, which is totally the same as other distribution matching methods. Accordingly, the accuracy should degrade in this case. According to Fig. 10.5b, as we increase t, the accuracies of OS and OS* decrease and the overall accuracy improves. This result indicates that the model does not learn representations effective to distinguish unknown examples from known ones.

Probability for Unknown Class In Fig. 10.5c and d, frequency diagram of the probability of an unknown category is presented in the adaptation from Webcam to DSLR. At the beginning of training, Fig. 10.5c, the probability is low in most samples including the known and unknown samples. As shown in Fig. 10.5d, many unknown samples have high probability of unknown class while many known samples have low probability of the unknown class after 500 epochs. We can observe that unknown and known samples are separated as training proceeds.

10.4.2.2 21 Class Classification

In addition, we will show the results when the dataset has larger number of classes. In addition to the 10 known classes used in previous experiments, the samples of 10 classes which were not used in the previous setting are added. The 10 classes are employed as unknown classes in the source domain in [4]. To summarize, we conducted 21 class classification experiments in this setting. We also evaluate OSBP on VGG Network. With regard to other details, we followed the protocol of the previous experiment.

The results are described in Table 10.2. The superiority of OSBP over baselines is clear. The effectiveness of MMD and BP is not observed in this setting too. The result of adaptation from Amazon to Webcam (A–W) reveals that the accuracy of other methods is better than our approach based on OS* and OS. However, their scores of "ALL" are much worse than OSBP. "ALL" indicates the accuracy measured on all the samples without averaging over classes. Thus, the consequence indicates that existing methods tend to recognize target samples as one of known classes in this setting. From the results, the effectiveness of OSBP is verified when the number of class increases.

10.4.3 Experiments on VisDA Dataset

In this section, we will present evaluation OSBP on adaptation from a synthetic to real dataset. VisDA dataset [27] has 12 categories in total. The source domain images are gathered by rendering 3D synthesized models whereas the target domain consists of real images. 6 categories (bicycle, bus, car, motorcycle, train and truck) were selected and other 6 categories were defined as the unknown class (aeroplane, horse, knife, person, plant and skateboard). The training procedure of the networks is the same as used for experiments on Office.

The results are presented in Table 10.3. OSBP performed better than the other baselines in most metric. *Avg* shows the accuracy averaged over all classes. *Avg known* means the accuracy averaged over only known classes. In both evaluation metrics, OSBP showed better performance, which indicates that OSBP is better both at aligning distributions between known classes and rejecting unknown class. We further show the examples of images and prediction of OSBP in Table 10.4. We show both failure case and successful case here. As we can observe, most images have multiple classes of objects or objects are somewhat occluded. In the second columns from the left, the images are classified as motorcycle though they are

Table 10.2 Accuracy (%) of experiments on Office dataset in 20 shared class situation

Adaptation Scenario	A–D			A–W			D–A		
	OS	OS*	ALL	OS	OS*	ALL	OS	OS*	ALL
OSVM	73.6 ± 0.4	**75.8** ± 0.6	57.6	**72.0** ± 0.5	**74.1** ± 0.5	58.0	44.9 ± 0.1	43.9 ± 0.1	51.1
MMD + OSVM	72.1 ± 0.9	73.9 ± 1.0	57.8	69.1 ± 0.8	71.2 ± 0.9	54.9	29.8 ± 0.6	26.5 ± 0.6	50.3
BP + OSVM	70.4 ± 0.2	72.1 ± 0.3	57.1	70.9 ± 0.5	72.9 ± 0.4	57.6	30.9 ± 0.2	27.6 ± 0.2	51.3
Ours	**74.8** ± 0.5	74.6 ± 0.5	**73.9**	66.8 ± 3.5	66.1 ± 3.7	**69.7**	**64.6** ± 1.2	**65.9** ± 4.9	**68.5**

Adaptation Scenario	D–W			W–A			W–D			AVG		
	OS	OS*	ALL	OS	OS*	ALL	OS	OS*	ALL	OS	OS*	ALL
OSVM	63.1 ± 1.1	61.9 ± 1.2	69.9	34.0 ± 0.9	31.8 ± 1.3	48.3	82.9 ± 1.7	82.9 ± 2.3	84.2	61.8	61.7	61.5
MMD + OSVM	58.3 ± 0.6	56.6 ± 0.6	68.8	39.7 ± 2.1	37.1 ± 2.4	55.9	84.2 ± 1.3	84.5 ± 1.2	87.2	58.9	58.2	62.3
BP + OSVM	63.2 ± 2.8	61.7 ± 3.0	71.3	40.0 ± 2.7	37.4 ± 3.0	56.0	83.1 ± 0.8	83.5 ± 0.8	86.4	59.8	59.1	63.2
Ours	**83.1** ± 0.6	**82.5** ± 0.6	**84.9**	**65.9** ± 0.1	**65.3** ± 0.2	**69.0**	**93.3** ± 0.2	**92.8** ± 0.2	**90.3**	**74.7**	**74.6**	**76.1**

We used VGG Network to obtain the results

Table 10.3 Accuracy (%) on VisDA dataset

Method	Bcycle	Bus	Car	Mcycle	Train	Truck	Unknown	Avg	Avg known
AlexNet									
OSVM	4.8	45.0	44.2	43.5	59.0	10.5	57.4	37.8	34.5
OSVM+MMD	0.2	30.9	49.1	54.8	56.1	8.1	61.3	37.2	33.2
OSVM+BP	9.1	50.5	**53.9**	79.8	69.0	8.1	42.5	44.7	45.1
Ours	**48.0**	**67.4**	39.2	**80.2**	69.4	24.9	80.3	**58.5**	**54.8**
VGGNet									
OSVM	31.7	51.6	66.5	70.4	**88.5**	20.8	38.0	52.5	54.9
OSVM+MMD	39.0	50.1	64.2	79.9	86.6	16.3	44.8	54.4	56.0
OSVM+BP	31.8	56.6	**71.7**	77.4	87.0	22.3	41.9	55.5	57.8
Ours	**53.6**	**72.0**	49.1	**80.8**	81.9	29.4	89.7	**65.2**	**61.1**

The accuracy per class is shown

Table 10.4 Examples of recognition results on VisDA dataset

Ground Truth Class → Predicted Class			
Known → Unknown ×	**Unknown → Known ×**	**Known → Known √**	**Unknown → Unknown √**
Train → Unknown	Unknown → Motorcycle	Truck → Truck	Unknown → Unknown
Motorcycle → Unknown	Unknown → Motorcycle	Bicycle → Bicycle	Unknown → Unknown
Car → Unknown	Unknown → Motorcycle	Motorcycle → Motorcycle	Unknown → Unknown

unknown. Persons often show up in the images of motorcycle and the appearance of the person and horse have similar features to such motorcycle images. In the third and fourth columns, we show successful cases. If the single object shows up, the classification is successful in most cases.

10.4.4   Experiments on Digits Dataset

We further show evaluation of OSBP on digits dataset. We employed three digits datasets, SVHN [26], USPS [20] and MNIST. In this evaluation, we conducted 3 scenarios in total, namely, adaptation from SVHN to MNIST, USPS to MNIST and MNIST to USPS. These are popular scenarios to evaluate methods in unsupervised domain adaptation. The numbers from 0 to 4 were selected as known classes while the other numbers were chosen as unknown. In this experiment, OSBP were compared with two baselines, OSVM, MMD combined with OSVM, and BP combined with OSVM. With regard to OSVM, we first trained the network with source known samples and extracted features using the network, then used the features to train OSVM. When training the network, we used Adam [18] with a learning rate 2.0×10^{-5}.

Adaptation from SVHN to MNIST   In this setting, all SVHN training samples with numbers in the range from 0 to 4 are used to train the network. With regard to the target samples, all samples in the training splits of MNIST are utilized.

Adaptation Between USPS and MNIST   When using the datasets as a source domain, all training samples with number from 0 to 4 are used. As for the target datasets, we used all training samples.

Result   The results are presented in Table 10.5. OSBP performed better than other baselines. Especially, with regard to the adaptation between USPS and MNIST, OSBP achieves very accurate recognition. In contrast, the performance of adaptation between SVHN and MNIST is worse than the adaptation between USPS and MNIST. There exists large domain difference between SVHN and MNIST, which causes the degrade in the performance. We also visualized the features extracted from the models in Fig. 10.6. Unknown classes (5~9) are separated by OSBP whereas known classes are aligned with source samples. The method based on distribution matching fails in adaptation for this open set scenario. We can see that BP attempts to align all of the target features with source features. As a result, unknown target samples are made hard to detect, which is clear from the qualitative results of BP. The accuracy of UNK in BP + OSVM is much worse than the others.

10.5   Conclusion and Future Work

In this chapter, we have shown the recent advance in domain adaptation where the source and target domain datasets do not completely share the categories. We focused on open set domain adaptation and introduced an adversarial learning method called OSBP. This method enables the generation of features that can separate unknown target samples from known target samples, which is definitely different from existing distribution matching methods. Moreover, this approach does

Table 10.5 Accuracy (%) of experiments on digits dataset

Method	SVHN-MNIST				USPS-MNIST				MNIST-USPS				Average			
	OS	OS*	ALL	UNK	OS	OS*	ALL	UNK	OS	OS*	ALL	UNK	OS	OS*	ALL	UNK
OSVM	54.3	63.1	37.4	10.5	43.1	32.3	63.5	97.5	79.8	77.9	84.2	**89.0**	59.1	57.7	61.7	65.7
MMD + OSVM	55.9	64.7	39.1	12.2	62.8	58.9	69.5	82.1	80.0	79.8	81.3	81.0	68.0	68.8	66.3	58.4
BP + OSVM	62.9	**75.3**	39.2	0.7	84.4	**92.4**	72.9	0.9	33.8	40.5	21.4	44.3	60.4	69.4	44.5	15.3
Ours	**63.0**	59.1	**71.0**	**82.3**	**92.3**	91.2	**94.4**	**97.6**	**92.1**	**94.9**	**88.1**	78.0	**82.4**	**81.7**	**84.5**	**85.9**

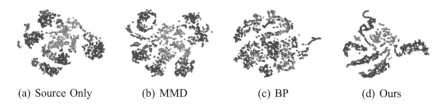

(a) Source Only   (b) MMD   (c) BP   (d) Ours

Fig. 10.6 Feature visualization of adaptation from USPS to MNIST. Visualization of source and target features. **Blue points** are source features. **Red points** are target known features. **Green points** are target unknown features. (**a**) Source only. (**b**) MMD. (**c**) BP. (**d**) Ours

not require unknown source samples. We have seen the effectiveness of this method through various experiments.

Recently, the research on open-set or partial domain adaptation is getting attention in domain adaptation research. However, to propose a method, we still utilize assumption that we know we have unknown classes in the target domain or do not have some of the known classes. As a way to tackle this problem, a task called universal domain adaptation is proposed [40]. This task tackles the situation where we do not have knowledge on the label distribution of target domain. This is a very hard task. However, considering the problem setting of domain adaptation, the target domain can be black-box and this setting is very practical. We believe that methods on this kind of tasks should be developed.

References

1. Bendale, A., Boult, T.E.: Towards open set deep networks. In: Proceedings of the IEEE Conference on Computer Vision and Pattern Recognition (2016)
2. Bousmalis, K., Trigeorgis, G., Silberman, N., Krishnan, D., Erhan, D.: Domain separation networks. In: Conference on Advances in Neural Information Processing Systems (2016)
3. Bousmalis, K., Silberman, N., Dohan, D., Erhan, D., Krishnan, D.: Unsupervised pixel-level domain adaptation with generative adversarial networks. In: Proceedings of the IEEE Conference on Computer Vision and Pattern Recognition (2017)
4. Busto, P.P., Gall, J.: Open set domain adaptation. In: Proceedings of the IEEE International Conference on Computer Vision (2017)
5. Cao, Z., Ma, L., Long, M., Wang, J.: Partial adversarial domain adaptation. In: Proceedings of the European Conference on Computer Vision (2018)
6. Chen, Y., Li, W., Sakaridis, C., Dai, D., Van Gool, L.: Domain adaptive faster r-cnn for object detection in the wild. In: Proceedings of the IEEE Conference on Computer Vision and Pattern Recognition (2018)
7. Deng, J., Dong, W., Socher, R., Li, L.J., Li, K., Fei-Fei, L.: Imagenet: A large-scale hierarchical image database. In: Proceedings of the IEEE Conference on Computer Vision and Pattern Recognition (2009)
8. Ganin, Y., Lempitsky, V.: Unsupervised domain adaptation by backpropagation. In: International Conference on Machine Learning (2015)
9. Ge, Z., Demyanov, S., Chen, Z., Garnavi, R.: Generative openmax for multi-class open set classification. In: British Machine Vision Conference (2017)

10. Ghifary, M., Kleijn, W.B., Zhang, M., Balduzzi, D., Li, W.: Deep reconstruction-classification networks for unsupervised domain adaptation. In: Proceedings of the European Conference on Computer Vision (2016)
11. Gong, B., Shi, Y., Sha, F., Grauman, K.: Geodesic flow kernel for unsupervised domain adaptation. In: Proceedings of the IEEE Conference on Computer Vision and Pattern Recognition (2012)
12. Gong, B., Grauman, K., Sha, F.: Connecting the dots with landmarks: Discriminatively learning domain-invariant features for unsupervised domain adaptation. In: International Conference on Machine Learning (2013)
13. Goodfellow, I., Pouget-Abadie, J., Mirza, M., Xu, B., Warde-Farley, D., Ozair, S., Courville, A., Bengio, Y.: Generative adversarial nets. In: Conference on Advances in Neural Information Processing Systems (2014)
14. Gretton, A., Borgwardt, K.M., Rasch, M., Schölkopf, B., Smola, A.J.: A kernel method for the two-sample-problem. In: Conference on Advances in Neural Information Processing Systems (2007)
15. Hoffman, J., Wang, D., Yu, F., Darrell, T.: Fcns in the wild: Pixel-level adversarial and constraint-based adaptation (2016). Preprint. arXiv:1612.02649
16. Ioffe, S., Szegedy, C.: Batch normalization: Accelerating deep network training by reducing internal covariate shift. In: International Conference on Machine Learning (2015)
17. Jain, L.P., Scheirer, W.J., Boult, T.E.: Multi-class open set recognition using probability of inclusion. In: Proceedings of the European Conference on Computer Vision (2014)
18. Kingma, D., Ba, J.: Adam: A method for stochastic optimization (2014). Preprint. arXiv:1412.6980
19. Krizhevsky, A., Sutskever, I., Hinton, G.E.: Imagenet classification with deep convolutional neural networks. In: Conference on Advances in Neural Information Processing Systems (2012)
20. LeCun, Y., Bottou, L., Bengio, Y., Haffner, P.: Gradient-based learning applied to document recognition. Proc. IEEE **86**(11), 2278–2324 (1998)
21. Liu, M.Y., Breuel, T., Kautz, J.: Unsupervised image-to-image translation networks. In: Conference on Advances in Neural Information Processing Systems (2017)
22. Long, M., Cao, Y., Wang, J., Jordan, M.I.: Learning transferable features with deep adaptation networks. In: International Conference on Machine Learning (2015)
23. Long, M., Zhu, H., Wang, J., Jordan, M.I.: Unsupervised domain adaptation with residual transfer networks. In: Conference on Advances in Neural Information Processing Systems (2016)
24. Long, M., Wang, J., Jordan, M.I.: Deep transfer learning with joint adaptation networks. In: International Conference on Machine Learning (2017)
25. Maaten, L.V.D., Hinton, G.: Visualizing data using t-SNE. J. Mach. Learn. Res. **9**(Nov), 2579–2605 (2008)
26. Netzer, Y., Wang, T., Coates, A., Bissacco, A., Wu, B., Ng, A.Y.: Reading digits in natural images with unsupervised feature learning. In: Neural Information Processing Systems Workshop on Deep Learning and Unsupervised Feature Learning (2011)
27. Peng, X., Usman, B., Kaushik, N., Hoffman, J., Wang, D., Saenko, K.: Visda: The visual domain adaptation challenge (2017). Preprint. arXiv:1710.06924
28. Saenko, K., Kulis, B., Fritz, M., Darrell, T.: Adapting visual category models to new domains. In: Proceedings of the European Conference on Computer Vision (2010)
29. Saito, K., Ushiku, Y., Harada, T.: Asymmetric tri-training for unsupervised domain adaptation. In: International Conference on Machine Learning (2017)
30. Saito, K., Watanabe, K., Ushiku, Y., Harada, T.: Maximum classifier discrepancy for unsupervised domain adaptation (2017). Preprint. arXiv:1712.02560
31. Saito, K., Yamamoto, S., Ushiku, Y., Harada, T.: Open set domain adaptation by backpropagation. In: Proceedings of the European Conference on Computer Vision (2018)

32. Saito, K., Ushiku, Y., Harada, T., Saenko, K.: Strong-weak distribution alignment for adaptive object detection. In: Proceedings of the IEEE Conference on Computer Vision and Pattern Recognition (2019)
33. Salimans, T., Goodfellow, I., Zaremba, W., Cheung, V., Radford, A., Chen, X.: Improved techniques for training gans. In: Conference on Advances in Neural Information Processing Systems (2016)
34. Sener, O., Song, H.O., Saxena, A., Savarese, S.: Learning transferrable representations for unsupervised domain adaptation. In: Conference on Advances in Neural Information Processing Systems (2016)
35. Simonyan, K., Zisserman, A.: Very deep convolutional networks for large-scale image recognition (2014). Preprint. arXiv:1409.1556
36. Taigman, Y., Polyak, A., Wolf, L.: Unsupervised cross-domain image generation. In: International Conference on Learning Representations (2016)
37. Tzeng, E., Hoffman, J., Zhang, N., Saenko, K., Darrell, T.: Deep domain confusion: Maximizing for domain invariance (2014). Preprint. arXiv:1412.3474
38. Tzeng, E., Hoffman, J., Saenko, K., Darrell, T.: Adversarial discriminative domain adaptation. In: Proceedings of the IEEE Conference on Computer Vision and Pattern Recognition (2017)
39. Yan, H., Ding, Y., Li, P., Wang, Q., Xu, Y., Zuo, W.: Mind the class weight bias: Weighted maximum mean discrepancy for unsupervised domain adaptation. In: Proceedings of the IEEE Conference on Computer Vision and Pattern Recognition (2017)
40. You, K., Long, M., Cao, Z., Wang, J., Jordan, M.I.: Universal domain adaptation. In: Proceedings of the IEEE Conference on Computer Vision and Pattern Recognition (2019)
41. Zhang, J., Ding, Z., Li, W., Ogunbona, P.: Importance weighted adversarial nets for partial domain adaptation. In: Proceedings of the IEEE Conference on Computer Vision and Pattern Recognition (2018)

Chapter 11
Universal Domain Adaptation

**Kaichao You, Mingsheng Long, Zhangjie Cao, Jianmin Wang,
and Michael I. Jordan**

11.1 Introduction

In the last decade, we have seen the power of deep learning in the area of computer
vision. State of the art results are renewed for many vision tasks including object
detection [16], image classification [8] and semantic segmentation [9]. However,
deep learning algorithms requires a large amount of labeled training data to achieve,
which means tedious labor work on data collecting and labeling. When we dive into
a new scenario with abundant unlabeled data, it is usually hard to get enough labels
to feed the data-hungry deep models such that they can generalize well on unseen
data. A promising way is to make the best of existing labeled data from a related
domain (denote as source domain) to assist the model get better performance in
the target domain (the domain of interest). Compared with the data in the source
domain, the data in the target domain may be collected from different angles,
by cameras with different brands and configurations, or under different lighting
conditions, leading to the **domain gap**. Intuitively, the domain gap can be closed
because it is easy for human learners. To overcome the domain gap, researchers
have proposed visual domain adaptation [18], attempting to transfer a model from
the source domain to the target domain.

Domain adaptation methods mentioned in previous chapters tackle the domain
gap either by learning domain invariant features to close the marginal distribution
discrepancy, by generating features/samples as well as the corresponding labels for
the target domain, or by translating samples between domains while preserving

K. You · M. Long (✉) · Z. Cao · J. Wang
School of Software, Tsinghua University, Beijing, China
e-mail: mingsheng@tsinghua.edu.cn

M. I. Jordan
University of California, Berkeley, CA, USA

© Springer Nature Switzerland AG 2020

H. Venkateswara, S. Panchanathan (eds.), *Domain Adaptation in Computer
Vision with Deep Learning*, https://doi.org/10.1007/978-3-030-45529-3_11

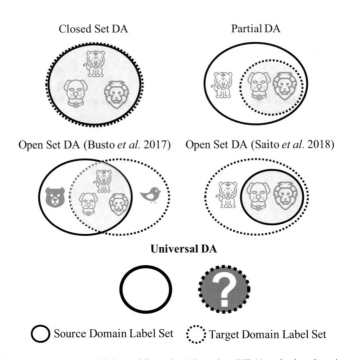

Fig. 11.1 Comparison among Universal Domain Adaptation (UDA) and other domain adaptation settings w.r.t the label set relationship between the source domain and the target domain (the common parts are shaded in blue). Only UDA deals with the setting where the target label set is unknown

the semantic information. Although novel methods are adopted, they only focus on the domain gap. An underlying assumption is that the source label set is identical to the target label set, as described in Fig. 11.1 (closed set domain adaptation). The simplified problem is elegant in math and enables researchers to focus on the core problem of domain adaptation. This line of research provides insights for future research. Recent works try to step out of the comfortable zone to relax the assumption, resulting in the emergence of open set domain adaptation [11, 14, 19] and partial domain adaptation [1, 3, 22]. As described in Fig. 11.1, partial domain adaptation [1, 22] relies on a weaker assumption that the target label set forms a subset of the source label set. Busto et al. [14] makes a step further by introducing "unknown" classes in both domains. However, they still assume it is known a priori which classes are common between two domains even in the training stage. A variant of open set domain adaptation proposed by Saito et al. [19] assumes that the target label set is a superset of the source label set. Luo et al. [13] claims the ability to solve domain adaptation problems with partly shared label sets, but requires a few labeled data in each target class, which means the structure of the target label set must be transparent. These works keep pushing domain adaptation towards a more practical setting.

Although researchers have been devoted to adding practical values for domain adaptation, none of these assumptions can hold for domain adaptation in the wild. Take recognition problem in the wild as an example: we are asked to recognize animals in a new area. Unlabeled data in the new area can be easily collected by setting up a camera in the wild. Labeled data in other areas would also be available. However, we cannot pick any of the methods in closed set DA, partial DA or open set DA, cause the label set of the new area is unknown and any assumption cannot be verified. If the source label set is large enough to contain the target label set, partial domain adaptation methods are good choices; if the source label set is contained in the target label set or common classes are known, open set domain adaptation methods are good choices. In a general scenario, however, we cannot select the proper domain adaptation method because no prior knowledge about the target domain label set is given.

To this end, it is clear that we need a more general domain adaptation setting. Here we study the setting proposed in [21] named **Universal Domain Adaptation (UDA)**. In UDA, we are given a labeled source domain, and the task is to classify the sample in the target domain correctly if it belongs to any class in the source label set or mark it as "unknown" otherwise. To make UDA more practical, we require that UDA should work for any target domain related to the given source domain, regardless of how its label set differs from that of the source domain. That is to say, models for UDA should work across a wide spectrum of different target domains. The word *"universal"* indicates that UDA imposes no prior knowledge on the label sets.

While we come up with a more general UDA setting, the challenges in UDA are harder. To design a model for the universal domain adaptation, we must solve two major technical challenges: (1) Since the target label set is unknown, somehow we have to automatically discover its structure, then we can decide which part of the source domain should be matched to which part of the target domain. If we simply match the entire source domain with the entire target domain, mismatching among different label sets will deteriorate the model. (2) The model should be equipped with the ability to reject a target sample (mark it as "unknown") if it does not belong to any class in the source label set. Since we only know classes in the source domain, it is impossible for the model to tell their detailed categories.

Next we study a specific solution termed **Universal Adaptation Network (UAN)** for the Universal Domain Adaptation. It carefully designs a novel criterion to quantify the transferability of each sample. The criterion is a sample-level weighting mechanism consisting of both the domain **similarity** and the prediction **uncertainty** of each sample. With the help of the transferability criterion, UAN can automatically detect and then match samples coming from the common label set between the source and target domains. At the same time, target samples from the target private label set can be correctly rejected as "unknown" class.

The main contents of this chapter are:

(1) A more practical Universal Domain Adaptation (UDA) setting with no restriction on the label set relation between the source label set and the target label set.

It pushes domain adaptation towards a nearly practical direction as target labels
are not accessible in unsupervised domain adaptation, not to mention how the
target label set overlaps with the source label set.

(2) A performance evaluation of existing domain adaptation methods in a wide
spectrum of UDA settings including closed set, partial and open set domain
adaptation. Methods with strong assumptions (and thus tailored to specific
settings) do not work well in UDA. This highlights the need for a UDA-friendly
model.

(3) A detailed description of the first UDA model named Universal Adaptation
Network (UAN). It is an end-to-end solution that can discover label sets
shared by both domains and promote common-class adaptation. Specifically,
it develops a weighting mechanism by using both the domain similarity and the
prediction uncertainty of each sample. Compared with existing methods, UAN
is experimentally verified to work stably and achieve state of the art results
across different UDA settings.

11.2 Related Work

Previous chapters can serve as related works for this chapter. For detailed related
works, please check the conference version [21].

11.3 Universal Domain Adaptation

In this section, Universal Domain Adaptation (UDA) setting is formally defined and
its solution (Universal Adaptation Network) is described in detail, from the origin
of ideas to the specific formula.

11.3.1 Problem Setup

In Universal Domain Adaptation (UDA), we are provided with a source domain
$\mathcal{D}_s = \{(\mathbf{x}_i^s, \mathbf{y}_i^s)\}$ consisting of n_s labeled samples and a target domain $\mathcal{D}_t = \{(\mathbf{x}_i^t)\}$
of n_t unlabeled samples. The underlying distribution of the source data is denoted
as p while the target distribution is called q. $\mathcal{C}_{sub}$ is used for the notation of label
sets, with its subscript denoting the specific meaning of the label set. $\mathcal{C}_s$ means
the label set of source domain and $\mathcal{C}_t$ means the label set of target domain. As
mentioned previously, the union of both label sets can be divided into three parts:
the overlap part $\mathcal{C} = \mathcal{C}_s \cap \mathcal{C}_t$ (labels common in both domains), the part private
to the source domain $\overline{\mathcal{C}}_s = \mathcal{C}_s \setminus \mathcal{C}$ (labels private to the source domain) and the
part private to the target domain $\overline{\mathcal{C}}_t = \mathcal{C}_t \setminus \mathcal{C}$ (labels private to the target domain).

p_{C_s} and p_C are shorthand notations to denote the underlying distributions of source data whose labels are in the label set C_s and C, and the same applies to q_{C_t}, q_C for target distributions. Keep in mind that the target examples are *fully unlabeled*, and the target label set C_t (inaccessible at training) is only used for defining the problem.

The **commonness** between two domains is defined as the Jaccard distance of two label sets: $\xi = \frac{|C_s \cap C_t|}{|C_s \cup C_t|}$. Now we can unify various domain adaptation settings: closed set domain adaptation is a special case of UDA when $\xi = 1$. The larger ξ is, the more sharing knowledge is and the less difficult the adaptation is. ξ partly reflects the feasibility of domain adaptation. **The task for UDA is to design a model that does not know ξ but works well across a wide spectrum of ξ.** It calls for the ability to tell target data in C from target data in $\overline{C}_t$, as well as to learn a classification model f to minimize the target error with common labels, i.e. $\min \mathbb{E}_{(\mathbf{x},\mathbf{y}) \sim q_C} [f(\mathbf{x}) \neq \mathbf{y}]$.

11.3.2   Technical Challenges

To solve the UDA problem, we have to understand its challenges. The very first challenge is the **category gap** between two domains. The cause of the category gap lies in the fact that the label sets across domains are different. Closed set domain adaptation methods work well when $C_s = C_t$, but they would wrongly match the source data in $\overline{C}_s$ with target data from $\overline{C}_t$ in UDA. Such blind alignment can cause problems since these two label sets have no overlap ($\overline{C}_s \cap \overline{C}_t = \emptyset$) and they shall not be matched in any way. Forcefully matching them will make target examples coming from $\overline{C}_t$ be predicted with a label in $\overline{C}_s$ whereas their ground truth label should be "unknown". Tailored methods in partial or open set domain adaptation have carefully designed mechanism to select the part of data to be matched. The fact that the relationship between C_s and C_t is unknown, however, stops us from selecting the proper algorithm to solve the problem. Thus, it is favorable if we can automatically select out the source and target data from C then confine feature alignment in the auto-discovered common label set.

The category gap is characteristic of UDA. Despite handling the category gap, we still have to deal with the **domain gap** in UDA, i.e. the distribution mismatch between the source and target distributions in the common label set. In other words, $p \neq q$ (especially $p_C \neq q_C$), a similar form we encounter in the standard closed set domain adaptation.

Another challenge arising from $\overline{C}_t$ is to **reject "unknown" classes**. In the literature, it is called "anomaly detection" and carried out by confidence thresholding with a pre-defined confidence threshold: each sample with classification confidence lower than the threshold is marked as "unknown". However, this straightforward implementation may well fail in universal domain adaptation because the predictions of modern neural networks are typically over-confident [7] but less discriminative due to the underlying domain gap.

11.3.3 Universal Adaptation Network

Now it's time for the nuts and bolts in the Universal Adaptation Network. As described in Fig. 11.2, there are several modules in UAN: a feature extractor F, a label predictor G, an adversarial domain discriminator D and a non-adversarial domain discriminator D'. The computation flow starts with the input $\mathbf{x}$ from either domain. Then $\mathbf{x}$ is fed into the feature extractor F. The label predictor G accepts the extracted feature $\mathbf{z} = F(\mathbf{x})$ to produce the probability $\hat{\mathbf{y}} = G(\mathbf{z})$ of $\mathbf{x}$ coming from the source classes C_s. The adversarial domain discriminator D aims to adversarially match the feature distributions of the source and target data falling in the common label set C (Note that we need a mechanism to detect the common label set). The non-adversarial domain discriminator D' obtains the **domain similarity** $\hat{d}' = D'(\mathbf{z})$, quantifying the similarity of $\mathbf{x}$ to the source domain. E_G, $E_{D'}$ and E_D represent the error for label predictor G, non-adversarial domain discriminator D' and adversarial domain discriminator D, which are formally defined as

$$E_G = \mathbb{E}_{(\mathbf{x},\mathbf{y})\sim p} L\left(\mathbf{y}, G(F(\mathbf{x}))\right) \tag{11.1}$$

$$E_{D'} = -\mathbb{E}_{\mathbf{x}\sim p} \log D'\left(F(\mathbf{x})\right)$$
$$\qquad - \mathbb{E}_{\mathbf{x}\sim q} \log\left(1 - D'\left(F(\mathbf{x})\right)\right) \tag{11.2}$$

$$E_D = -\mathbb{E}_{\mathbf{x}\sim p} w^s(\mathbf{x}) \log D\left(F(\mathbf{x})\right)$$
$$\qquad - \mathbb{E}_{\mathbf{x}\sim q} w^t(\mathbf{x}) \log\left(1 - D\left(F(\mathbf{x})\right)\right) \tag{11.3}$$

where L is the standard cross-entropy loss for classification, and $w^s(\mathbf{x})$ and $w^t(\mathbf{x})$ indicate the probability of a source/target sample belonging to the common label set C. The construction details of $w^s(\mathbf{x})$ and $w^t(\mathbf{x})$ will be explained in the next

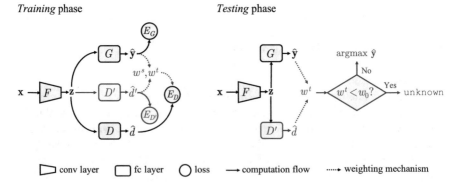

Fig. 11.2 The training and testing phases of the Universal Adaptation Network (UAN) described in this chapter for Universal Domain Adaptation (UDA)

subsection. The nice property of Eq. (11.3) is that D can be confined to distinguish the source and target data in the common label set C when $w^s(\mathbf{x})$ and $w^t(\mathbf{x})$ are well-established. Through an adversarial min-max game, F and D work together to achieve domain-invariant features in the common label set C. The label predictor G trained on such features can be safely transferred to the target domain.

The training process of UAN can be formulated as a minimax game:

$$\max_{D} \min_{F,G} E_G - \lambda E_D$$
$$\min_{D'} E_{D'}$$
(11.4)

where λ balances transferability and discriminability. By inserting the gradient reversal layer [4] between F and D, all the modules can be optimized in an end-to-end training fashion.

The testing process of UAN is described in the right sub-plot in Fig. 11.2. For every input test sample $\mathbf{x}$, we combine its categorical prediction $\hat{\mathbf{y}}(\mathbf{x})$ and the domain prediction $\hat{d}'(\mathbf{x})$ to compute $w^t(\mathbf{x})$ using Eq. (11.8) (details in the next subsection). With a hyperparameter threshold w_0 which controls the transferability, the prediction $y(\mathbf{x})$ can be written as

$$y(\mathbf{x}) = \begin{cases} \text{unknown} & w^t < w_0 \\ \text{argmax}\left(\hat{\mathbf{y}}\right) & w^t \geq w_0 \end{cases}$$
(11.5)

which either classifies it to one of the source classes or rejects the target sample as "unknown" class.

11.3.4   Transferability Criterion

In this section, we dive into the details about the sample-level transferability criterion $w^s = w^s(\mathbf{x})$ and $w^t = w^t(\mathbf{x})$. As we have analyzed before, a good sample-level transferability criterion is necessary. This way, each sample in both domains can be weighted during the adversarial training such that distribution alignment only happens for source and target data in the common label set C. Another usage of the transferability criterion is to assist the process of "unknown" sample detection which detects data from target private label set $\overline{C}_t$. Ideally, a well-established sample-level transferability criterion should satisfy the inequalities in Eq. (11.6):

$$\mathbb{E}_{\mathbf{x} \sim p_C} w^s(\mathbf{x}) > \mathbb{E}_{\mathbf{x} \sim p_{\overline{C}_s}} w^s(\mathbf{x})$$
$$\mathbb{E}_{\mathbf{x} \sim q_C} w^t(\mathbf{x}) > \mathbb{E}_{\mathbf{x} \sim q_{\overline{C}_t}} w^t(\mathbf{x})$$
(11.6)

These inequalities should hold in a large margin.

So the last question is how to construct the sample-level transferability criterion. First, let's list all the information we know about each input $\mathbf{x}$: $\hat{y}$, $\hat{d}$, $\hat{d}'$. Since D participates in the adversarial training and thus fooled, its output $\hat{d}$ is not discriminative enough. It makes us turn to $\hat{y}$ and $\hat{d}'$.

Domain Similarity In Eq. (11.2), the goal of D' is to distinguish samples from source domain and samples from target domain. Thus, $\hat{d}'$ quantifies the domain similarity of each sample. A source sample with smaller $\hat{d}'$ means that it looks more similar to the target domain; a target sample with larger $\hat{d}'$ means that it looks more similar to the source domain. Therefore, we can hypothesize that $\mathbb{E}_{\mathbf{x} \sim p_{\overline{C}_s}} \hat{d}' > \mathbb{E}_{\mathbf{x} \sim p_C} \hat{d}' > \mathbb{E}_{\mathbf{x} \sim q_C} \hat{d}' > \mathbb{E}_{\mathbf{x} \sim q_{\overline{C}_t}} \hat{d}'$.

Here is an intuitive verification of the hypothesize: (1) Due to the nature of D', inequality $\mathbb{E}_{\mathbf{x} \sim p_{\overline{C}_s}} \hat{d}'$, $\mathbb{E}_{\mathbf{x} \sim p_C} \hat{d}' > \mathbb{E}_{\mathbf{x} \sim q_C} \hat{d}'$, $\mathbb{E}_{\mathbf{x} \sim q_{\overline{C}_t}} \hat{d}'$ naturally holds. (2) Since p_C and q_C share the same label set, p_C is closer to q_C compared with $q_{\overline{C}_t}$, and it is reasonable to hypothesize $\mathbb{E}_{\mathbf{x} \sim p_{\overline{C}_s}} \hat{d}' > \mathbb{E}_{\mathbf{x} \sim p_C} \hat{d}'$. The same observation applies to $\mathbb{E}_{\mathbf{x} \sim q_C} \hat{d}' > \mathbb{E}_{\mathbf{x} \sim q_{\overline{C}_t}} \hat{d}'$.

Prediction Uncertainty The discriminative information about the input is reflected directly in the prediction $\hat{y}$, but it is only reliable in the source domain because source domain is fully labeled. To exploit unlabeled data, entropy minimization has been used as a criterion in semi-supervised learning and domain adaptation [6, 12] to enforce the decision boundary in the unlabeled data not to pass through high-density area. In essence, entropy quantifies the uncertainty of prediction: larger entropy means less confident prediction. We hypothesize: $\mathbb{E}_{\mathbf{x} \sim q_{\overline{C}_t}} H(\hat{y}) > \mathbb{E}_{\mathbf{x} \sim q_C} H(\hat{y}) > \mathbb{E}_{\mathbf{x} \sim p_C} H(\hat{y}) > \mathbb{E}_{\mathbf{x} \sim p_{\overline{C}_s}} H(\hat{y})$.

An intuitive verification is also available: (1) Since source domain data are fully labeled and target domain data are totally unlabeled, predictions are certain for source samples (thanks to supervised learning) and uncertain for target samples, $\mathbb{E}_{\mathbf{x} \sim q_{\overline{C}_t}} H(\hat{y})$, $\mathbb{E}_{\mathbf{x} \sim q_C} H(\hat{y}) > \mathbb{E}_{\mathbf{x} \sim p_C} H(\hat{y})$, $\mathbb{E}_{\mathbf{x} \sim p_{\overline{C}_s}} H(\hat{y})$. (2) Similar samples from q_C and p_C may have an influence on each other. That is to say, the entropy of samples from p_C becomes larger because they can be influenced by the high entropy samples from q_C. Still, as $\overline{C}_s$ has no intersection with C_t, samples from $p_{\overline{C}_s}$ cannot be influenced by the target data and keep their highest certainty. So it's reasonable to hypothesize that $\mathbb{E}_{\mathbf{x} \sim p_C} H(\hat{y}) > \mathbb{E}_{\mathbf{x} \sim p_{\overline{C}_s}} H(\hat{y})$. Similarly, $\overline{C}_t$ has no intersection with C_s (data from $q_{\overline{C}_t}$ does not belong to any class in C_s), and thus the hypothesis $\mathbb{E}_{\mathbf{x} \sim q_{C_t}} H(\hat{y}) > \mathbb{E}_{\mathbf{x} \sim q_{\overline{C}}} H(\hat{y})$ is reasonable.

After the above analysis, we can compose the sample-level transferability criterion for source data examples and target data examples respectively as Eqs. (11.7) and (11.8):

$$w^s(\mathbf{x}) = \frac{H(\hat{y})}{\log |C_s|} - \hat{d}'(\mathbf{x}) \tag{11.7}$$

$$w^t(\mathbf{x}) = \hat{d}'(\mathbf{x}) - \frac{H(\hat{\mathbf{y}})}{\log|\mathcal{C}_s|} \tag{11.8}$$

A small trick is that the entropy is divided by its maximum value ($\log|\mathcal{C}_s|$) to be restricted in the range of [0, 1]. This way, it is comparable to the domain similarity measure $\hat{d}'$ who lies in the same range.

The universal adaptation network (UAN) introduced in this chapter utilizes the sample-level transferability criterion to identify and then separate source data in $\mathcal{C}$, $\overline{\mathcal{C}}_s$ and target data in $\mathcal{C}, \overline{\mathcal{C}}_t$. As such, the category gap is reduced. The domain gap is reduced as well by aligning features between domains in shared label set $\mathcal{C}$.

11.4 Experiments

For universal domain adaptation, there are 3 degrees of freedom on label sets: ξ, $|\mathcal{C}|$ and $|\overline{\mathcal{C}}_t|$. To perform a thorough evaluation, we compare UAN with state of the art methods tailored to various label set assumptions under numerous UDA settings on several standard datasets with different ξ, $|\mathcal{C}_s \cup \mathcal{C}_t|$, $\overline{\mathcal{C}}_t$ and $\overline{\mathcal{C}}_s$. Then, we conduct controlled experiments to explore the impact of ξ, $|\mathcal{C}_s \cup \mathcal{C}_t|$, $\overline{\mathcal{C}}_t$ and $\overline{\mathcal{C}}_s$ respectively. Lastly, analysis of the quality of sample-level transferability criterion and the hyperparameter sensitivity about the UAN model is presented. Code and data is available at https://github.com/thuml/Universal-Domain-Adaptation.

11.4.1 Experimental Setup

This subsection is about the datasets, evaluation protocols and implementation details.

11.4.1.1 Datasets

Office-31 [18] is a standard domain adaptation dataset with 31 categories in 3 visually distinct domains (**A, D, W**). The 10 classes shared by **Office-31** and **Caltech-256** [5] are selected as the common label set $\mathcal{C}$. Then in alphabetical order, the next 10 classes are used as the $\overline{\mathcal{C}}_s$, and the rest 11 classes are used as the $\overline{\mathcal{C}}_t$. Here $\xi = 0.32$.

Office-Home [20] is a larger dataset (compared with **Office-31**) with 65 classes in 4 different domains: Artistic images (**Ar**), Clip-Art images (**Cl**), Product images (**Pr**) and Real-World images (**Rw**). In alphabetical order, we use the first 10 classes as $\mathcal{C}$, the next 5 classes as $\overline{\mathcal{C}}_s$ and the rest as $\overline{\mathcal{C}}_t$. Here $\xi = 0.15$.

VisDA2017 [15] dataset focuses on the simulation to real setting. The source data consist of labeled images rendered by game engines and target data consist of unlabeled real-world images. Each domain shares the same 12 classes. We use the first 6 classes as $\mathcal{C}$, the next 3 classes as $\overline{\mathcal{C}}_s$ and the rest as $\overline{\mathcal{C}}_t$. Here $\xi = 0.50$.

ImageNet-Caltech is constructed from **ImageNet-1K** [17] and **Caltech-256**. Following the tradition of [1, 2], the 84 common classes shared by both domains are selected as the common label set $\mathcal{C}$ and their private classes are selected as the private label set respectively. This setting naturally falls into the UDA paradigm: since the target domain label set is unavailable, it may well overlap with the source label set but hold a private part as well. Two UDA tasks are: $\mathbf{I} \rightarrow \mathbf{C}$ and $\mathbf{C} \rightarrow \mathbf{I}$. Here $\xi = 0.07$.

These UDA settings are designed to be maximally consistent with the existing configurations [1, 2, 14, 19] and cover as many commonness levels ξ as possible. Brute-force enumeration of all possible combination for ξ, $\overline{\mathcal{C}}_t$ and $\overline{\mathcal{C}}_s$ is unacceptable.

11.4.1.2 Evaluation Details

Compared Methods UAN is compared with the following methods: (1) Convolutional Neural Network: **ResNet** [8] (without any adaptation), (2) close-set domain adaptation methods: Domain-Adversarial Neural Networks (**DANN**) [4], Residual Transfer Networks (**RTN**) [12], (3) partial domain adaptation methods: Importance Weighted Adversarial Nets (**IWAN**) [22], Partial Adversarial Domain Adaptation (**PADA**) [2], (4) open set domain adaptation methods: Assign-and-Transform-Iteratively (**ATI**) [14] (ATI-λ is compared and λ is derived as described in [14]), Open Set Back-Propagation (**OSBP**) [19]. Aforementioned methods are state of the art in their respective settings. It shall be valuable to study the performance of these methods in the practical UDA setting.

Evaluation Protocols The performance metric is the same as that in Visual Domain Adaptation (VisDA2018) Open-Set Classification Challenge, where all the classes in the target private label set are unified as an "unknown" class and the final metric is the average of per-class accuracy for all the $|\mathcal{C}| + 1$ classes. For methods without the ability of anomaly detection, we extend them by confidence thresholding: at the test stage, the input image is classified as "unknown" if the prediction confidence is under the confidence threshold.

Implementation Details The framework we choose is PyTorch. The backbone network (architecture of F and its initialization) is ResNet-50 [8]. Feature extractors are initialized from ResNet-50 pre-trained on ImageNet and then fine-tuned. When calculating $\hat{\mathbf{y}}$ in Eq. (11.7), we set temperature [10] as 10 because the prediction for source data is usually too certain and the entropy is low. When applied in Eq. (11.3), w^s, w^t are normalized in a mini-batch to be within interval [0, 1] so that weights in Eq. (11.3) are non-negative.

11.4.2 Classification Results

The main results can be found in Tables 11.1 and 11.2. In these experiments, UAN is empirically verified to be a proper solution for the UDA problem. Besides, there are several phenomena worth discussing.

(1) ResNet seems to be a strong baseline in universal domain adaptation. Many tailored methods struggle to catch up with ResNet, stressing the difficulty of UDA. In fact, many domain adaptation methods tend to suffer from negative transfer, the term used to describe worse performance than a model only trained on source data without any adaptation. Figure 11.4a quantifies the negative transfer by the per-class accuracy improvement over ResNet in the Office-Home task **Ar $\rightarrow$ Cl**. It shows DANN, IWAN, and OSBP suffer from negative transfer in most categories and they only promote the adaptation for a few categories. Only UAN achieves positive transfer among all classes thanks to the carefully designed sample-level transferability criterion. It discards data coming from $\overline{C}_t$ and $\overline{C}_s$ during feature alignment and yields a better criterion for detecting the "unknown" class than the naive confidence thresholding.

(2) Methods for other domain adaptation settings except UDA perform well when their assumptions hold but worse when violated. For example, OSBP assumes no source private classes. If source private classes are manually removed (invalid operation in the absence of target labels), the accuracy of OSBP is 89.1% on Office-31; however, if source private classes remains (violating its assumption), its accuracy drops dramatically to 67.68%. As the assumptions of compared methods are violated in UDA, it is no wonder that their accuracies drop sharply.

11.4.3 Analysis on Different UDA Settings

Varying Size of $\overline{C}_t$ and $\overline{C}_s$ Here we fix $|C_s \cup C_t|$ and ξ, to explore the results of these methods on universal domain adaptation with the different sizes of $\overline{C}_t$ ($\overline{C}_s$ changes simultaneously) on task **A $\rightarrow$ D** in Office-31. As described in Fig. 11.3a, UAN gets the best results on most sizes of $\overline{C}_t$. What's more, when $|\overline{C}_t| = 0$, satisfying the assumption of partial domain adaptation $C_t \subset C_s$, UAN is comparable to IWAN. And when $|\overline{C}_t| = 21$, satisfying the assumption of open set domain adaptation $C_s \subset C_t$, UAN is comparable to OSBP. IWAN and OSBP both take advantage of the prior knowledge about label sets and design modules to exploit the knowledge. However, UAN can still catch up with them in their expert settings, indicating UAN is effective and robust to diverse sizes of $\overline{C}_t$ and $\overline{C}_s$. When $0 < \overline{C}_t < 21$, where C_s and C_t partly overlap, UAN outperforms other methods with large margin. UAN can yield impressive results with no prior knowledge about the target label set. In Fig. 11.3a,

Table 11.1 Average class accuracy (%) of universal domain adaptation tasks on **Office-Home** ($\xi = 0.15$) dataset (ResNet)

Method	Office-Home												
	Ar→Cl	Ar→Pr	Ar→Rw	Cl→Ar	Cl→Pr	Cl→Rw	Pr→Ar	Pr→Cl	Pr→Rw	Rw→Ar	Rw→Cl	Rw→Pr	Avg
ResNet [8]	59.37	76.58	87.48	69.86	71.11	81.66	73.72	56.30	86.07	78.68	59.22	78.59	73.22
DANN [4]	56.17	81.72	86.87	68.67	73.38	83.76	69.92	56.84	85.80	79.41	57.26	78.26	73.17
RTN [12]	50.46	77.80	86.90	65.12	73.40	85.07	67.86	45.23	85.50	79.20	55.55	78.79	70.91
IWAN [22]	52.55	81.40	86.51	70.58	70.99	85.29	74.88	57.33	85.07	77.48	59.65	78.91	73.39
PADA [22]	39.58	69.37	76.26	62.57	67.39	77.47	48.39	35.79	79.60	75.94	44.50	78.10	62.91
ATI [14]	52.90	80.37	85.91	71.08	72.41	84.39	74.28	57.84	85.61	76.06	60.17	78.42	73.29
OSBP [19]	47.75	60.90	76.78	59.23	61.58	74.33	61.67	44.50	79.31	70.59	54.95	75.18	63.90
UAN w/o d	61.60	81.86	87.67	74.52	73.59	84.88	73.65	57.37	86.61	81.58	62.15	79.14	75.39
UAN w/o y	56.63	77.51	87.61	71.96	69.08	83.18	71.40	56.10	84.24	79.27	60.59	78.35	72.91
UAN	**63.00**	**82.83**	**87.85**	**76.88**	**78.70**	**85.36**	**78.22**	**58.59**	**86.80**	**83.37**	**63.17**	**79.43**	**77.02**

Bold numbers indicate the best results

Table 11.2 Average class accuracy (%) on **Office-31** ($\xi = 0.32$), **ImageNet-Caltech** ($\xi = 0.07$) and **VisDA2017** ($\xi = 0.50$) (ResNet)

Method	Office-31							ImageNet-Caltech		VisDA
	A → W	D → W	W → D	A → D	D → A	W → A	Avg	I → C	C → I	
ResNet [8]	75.94	89.60	90.91	80.45	78.83	81.42	82.86	70.28	65.14	52.80
DANN [4]	80.65	80.94	88.07	82.67	74.82	83.54	81.78	71.37	66.54	52.94
RTN [12]	85.70	87.80	88.91	82.69	74.64	83.26	84.18	71.94	66.15	53.92
IWAN [22]	85.25	90.09	90.00	84.27	84.22	86.25	86.68	72.19	66.48	58.72
PADA [22]	85.37	79.26	90.91	81.68	55.32	82.61	79.19	65.47	58.73	44.98
ATI [14]	79.38	92.60	90.08	84.40	78.85	81.57	84.48	71.59	67.36	54.81
OSBP [19]	66.13	73.57	85.62	72.92	47.35	60.48	67.68	62.08	55.48	30.26
UAN	**85.62**	**94.77**	**97.99**	**86.50**	**85.45**	**85.12**	**89.24**	**75.28**	**70.17**	**60.83**

Bold numbers indicate the best results

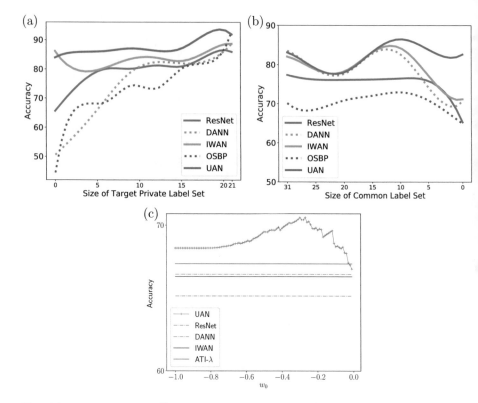

Fig. 11.3 (a) Accuracy w.r.t. $|\overline{\mathcal{C}}_t|$ in task **A → D**, $\xi = 0.32$. (b) Accuracy w.r.t. $|\mathcal{C}|$ in task **A → D**. (c) Performance w.r.t. threshold w_0

almost all methods get better when $|\overline{\mathcal{C}}_t|$ gets larger. This can be expected as larger $|\overline{\mathcal{C}}_t|$ indicates smaller $|\overline{\mathcal{C}}_s|$ and less distraction to the label predictor.

Varying Size of Common Label Set $\mathcal{C}$ In Office-31 dataset, $|\mathcal{C}| + |\overline{\mathcal{C}}_t| + |\overline{\mathcal{C}}_s| = 31$. To explore another dimension of UDA, we vary the size of $\mathcal{C}$. We set $|\overline{\mathcal{C}}_t| = |\overline{\mathcal{C}}_s| + 1$ for simplicity. $|\mathcal{C}|$ varies from 0 to 31. Figure 11.3b shows the performance of these methods under different sizes of $\mathcal{C}$. When $|\mathcal{C}| = 0$, i.e. $\mathcal{C}_t \cap \mathcal{C}_s = \emptyset$ (source domain and target domain are separate and have no overlap), UAN performs substantially better than all the other methods. This may be attributed to the fact that they wrongly assume some common labels across domains and thus cannot filter out "unknown" samples well. When $|\mathcal{C}| = 31$, where UDA degenerates to closed set DA with $\mathcal{C}_s = \mathcal{C}_t$, it's surprising UAN is still comparable with DANN, indicating that the sample-level transferability criterion of UAN won't hurt useful samples for adaptation. When $|\mathcal{C}|$ decreases, DANN and IWAN struggles but UAN still works stably.

11.4.4 Analysis of Universal Adaptation Network

Ablation Study We dive deeper into the root of the sample-level transferability criterion through the ablation study about variants of UAN: (1) **UAN w/o d** only integrates the uncertainty into the sample-level transferability criterion in Eqs. (11.7) and (11.8); (2) **UAN w/o y** only integrates the domain similarity into sample-level transferability criterion in Eqs. (11.7) and (11.8). Results for the ablation study can be found in the last rows of Table 11.1. Compared with UAN, UAN w/o d and UAN w/o y are worse, because they miss either the domain similarity component or the uncertainty criterion component in the definition of $w^s(\mathbf{x})$, $w^t(\mathbf{x})$. After a careful look, we find that UAN w/o d performs better than UAN w/o y, meaning that the uncertainty criterion is more crucial than the domain similarity in the sample-level transferability criterion.

Hypotheses Justification In Sect. 11.3.4, the hypotheses are intuitively verified. Here we verify the hypotheses empirically. The estimated probability density function for different components in Eqs. (11.7) and (11.8) are plotted in Fig. 11.4b. The plot shows that hypotheses in Sect. 11.3.4 are experimentally verified, a reason why UAN can perform well in various UDA settings. The uncertainty and the domain similarity across domains differ with each other. By combining these two components, the transferability criterion can be more distinguishable and useful for UAN.

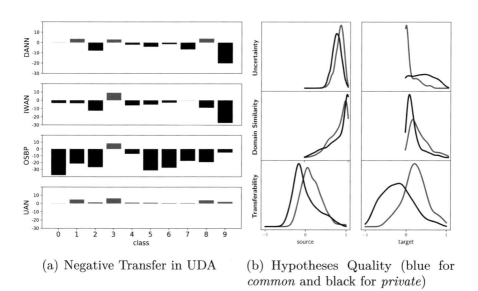

(a) Negative Transfer in UDA

(b) Hypotheses Quality (blue for *common* and black for *private*)

Fig. 11.4 (a) The negative transfer influence in UDA (task **Ar → Cl**). (b) Justification of validity of hypotheses in Sect. 11.3.4

Threshold Sensitivity UAN requires a pre-defined threshold w_0 to detect the "unknown" classes. We measure the sensitivity of UAN w.r.t threshold w_0 in task **I** $\rightarrow$ **C**. As described in Fig. 11.3c, UAN's accuracies depend on w_0 and can vary by about 2%. However, it outperforms the other methods by large margins for a wide range of w_0 even though other methods are fully tuned to get the best accuracies they can achieve.

11.5 Conclusion

In this chapter, we introduce a practical domain adaptation setting termed UDA to dispose of the scenario when we have unlabeled data and know nothing about the target label set. Together with the UDA setting, we introduce a solution named UAN from its idea to its implementation. UAN has a well-designed sample-level transferability criterion in order to address UDA. Experiments show that methods with the help of prior knowledge on the label set relationship will fail in the general UDA setting and only UAN works stably.

UAN can serve as a pilot study when we encounter a new domain adaptation scenario. If one wants to deploy a model to a new scenario, UAN can be the first choice to try. If most examples as marked as "unknown" by UAN, then it is impossible to expect successful adaptation, and collecting labels will be necessary. On the contrary, if UAN can generate labels for most examples, we can turn to other tailored domain adaptation methods and expect domain adaptation to work.

References

1. Cao, Z., Long, M., Wang, J., Jordan, M.I.: Partial transfer learning with selective adversarial networks. In: Proceedings of the IEEE Conference on Computer Vision and Pattern Recognition (2018)
2. Cao, Z., Ma, L., Long, M., Wang, J.: Partial adversarial domain adaptation. In: Proceedings of the European Conference on Computer Vision, pp. 135–150 (2018)
3. Cao, Z., You, K., Long, M., Wang, J., Yang, Q.: Learning to transfer examples for partial domain adaptation. In: Proceedings of the IEEE Conference on Computer Vision and Pattern Recognition (2019)
4. Ganin, Y., Ustinova, E., Ajakan, H., Germain, P., Larochelle, H., Laviolette, F., Marchand, M., Lempitsky, V.S.: Domain-adversarial training of neural networks. J. Mach. Learn. Res. **17**, 59:1–59:35 (2016)
5. Gong, B., Shi, Y., Sha, F., Grauman, K.: Geodesic flow kernel for unsupervised domain adaptation. In: Proceedings of the IEEE Conference on Computer Vision and Pattern Recognition (2012)
6. Grandvalet, Y., Bengio, Y.: Semi-supervised learning by entropy minimization. In: Conference on Advances in Neural Information Processing Systems, pp. 529–536 (2004)
7. Guo, C., Pleiss, G., Sun, Y., Weinberger, K.Q.: On calibration of modern neural networks. In: International Conference on Machine Learning, pp. 1321–1330

8. He, K., Zhang, X., Ren, S., Sun, J.: Deep residual learning for image recognition. In: Proceedings of the IEEE Conference on Computer Vision and Pattern Recognition (2016)
9. He, K., Gkioxari, G., Dollár, P., Girshick, R.: Mask R-CNN. In: International Conference on Computer Vision, pp. 2980–2988. IEEE (2017)
10. Hinton, G., Vinyals, O., Dean, J.: Distilling the knowledge in a neural network. In: NIPS Deep Learning and Representation Learning Workshop (2015)
11. Liu, H., Cao, Z., Long, M., Wang, J., Yang, Q.: Separate to adapt: Open set domain adaptation via progressive separation. In: Proceedings of the IEEE Conference on Computer Vision and Pattern Recognition (2019)
12. Long, M., Zhu, H., Wang, J., Jordan, M.I.: Unsupervised domain adaptation with residual transfer networks. In: Conference on Advances in Neural Information Processing Systems, pp. 136–144 (2016)
13. Luo, Z., Zou, Y., Hoffman, J., Fei-Fei, L.F.: Label efficient learning of transferable representations acrosss domains and tasks. In: Conference on Advances in Neural Information Processing Systems, pp. 165–177 (2017)
14. Panareda Busto, P., Gall, J.: Open set domain adaptation. In: International Conference on Computer Vision (2017)
15. Peng, X., Usman, B., Kaushik, N., Wang, D., Hoffman, J., Saenko, K., Roynard, X., Deschaud, J.E., Goulette, F., Hayes, T.L.: VisDA: A synthetic-to-real benchmark for visual domain adaptation. In: Proceedings of the IEEE Conference on Computer Vision and Pattern Recognition Workshops, pp. 2021–2026
16. Ren, S., He, K., Girshick, R., Sun, J.: Faster r-cnn: Towards real-time object detection with region proposal networks. In: Conference on Advances in Neural Information Processing Systems, pp. 91–99
17. Russakovsky, O., Deng, J., Su, H., Krause, J., Satheesh, S., Ma, S., Huang, Z., Karpathy, A., Khosla, A., Bernstein, M., Berg, A.C., Fei-Fei, L.: ImageNet Large Scale Visual Recognition Challenge. Int. J. Comput. Vision **115**(3), 211–252 (2015). https://doi.org/10.1007/s11263-015-0816-y
18. Saenko, K., Kulis, B., Fritz, M., Darrell, T.: Adapting visual category models to new domains. In: Proceedings of the European Conference on Computer Vision (2010)
19. Saito, K., Yamamoto, S., Ushiku, Y., Harada, T.: Open set domain adaptation by backpropagation. In: Proceedings of the European Conference on Computer Vision (2018)
20. Venkateswara, H., Eusebio, J., Chakraborty, S., Panchanathan, S.: Deep hashing network for unsupervised domain adaptation. In: Proceedings of the IEEE Conference on Computer Vision and Pattern Recognition (2017)
21. You, K., Long, M., Cao, Z., Wang, J., Jordan, M.I.: Universal domain adaptation. In: Proceedings of the IEEE Conference on Computer Vision and Pattern Recognition (2019)
22. Zhang, J., Ding, Z., Li, W., Ogunbona, P.: Importance weighted adversarial nets for partial domain adaptation. In: Proceedings of the IEEE Conference on Computer Vision and Pattern Recognition (2018)

Chapter 12
Multi-Source Domain Adaptation by Deep CockTail Networks

Ziliang Chen and Liang Lin

12.1 Introduction

Recent developments in deep representation learning have significantly improved state of the arts across a large variety of computer vision problems [15, 17, 26, 31]. These eyeball-catching advances, to a great extent, should be attributed to the availability of large scale datasets [3, 14] that provide an extensive array of labeled examples for supervised learning. However, when facing up with (unsupervised) Domain Adaptation [24] (**DA**), we are solely provided with labeled examples in a source domain while required to achieve the tasks without target labeled data (our work mainly concerns visual classification on the target domain). A straightforward solution is to directly learn a model with labeled source examples and then, deploy it to classify the target examples. But due to the maintenance of *domain shift* [11] between source and target domains, the performance of the source classifier almost degrades heavily on classifying target instances. To mitigate this damage, transferable representations are learned by minimizing source-target distribution discrepancy, endowing the classifier with the consistent classification ability on source and target examples [2, 7, 12].

It is worth noting that, existing DA algorithms are almost preconditioned on a single source where labeled examples are identically drawn from the source underlying distribution. This setup is widely-admitted for developing new DA approaches, though merely reflects a tip of the iceberg of realistic transfer circumstances. In many real-world cases, it is likely that we simultaneously witness multiple source domains. For instance, when training an object recognition module embedded in a household robot, people are allowed to exploit the labeled object images either from

Z. Chen (✉) · L. Lin
Sun Yat-sen University, Guangzhou, People's Republic of China
e-mail: linliang@ieee.org

© Springer Nature Switzerland AG 2020   213
H. Venkateswara, S. Panchanathan (eds.), *Domain Adaptation in Computer Vision with Deep Learning*, https://doi.org/10.1007/978-3-030-45529-3_12

Amazon.com (Source 1) or Flickr (Source 2). Another case regularly occurs in the area of medical imaging data collection. For the sake of typicality about sampled cases of illness, medical images are conventionally collected from hospitals all over the country in a long time.

These application circumstances produce a large amount of datasets that should be treated as a set of multiple sources. Consequently, the multi-source unsupervised domain adaptation (**MSDA**) problem has increasingly grabbed considerable attention in many applications [4, 13, 33], since reasonable approaches might bring about more transfer learning performance gains.

However, compared with witnessed progresses in deep single-source DA, scarce researches have paid attention to deep MSDA in terms of more complex domain-shift conditions. On one hand, the domain shifts among multiple sources indicates that, directly applying UDA by combining multiple sources as a single domain, is inefficient to address MSDA. On the other hand, based upon Liebig's law of the minimum, sources away from the target domain would more possibly threaten the target model performance. It might be helpful to mitigate this negative effect through filtering out those black-sheep sources, whereas hardly available for implementation: the black-sheep sources practically cab not be detected by human labor. An alternative is to balance the adaptations from multiple sources, but the balance principle is crucial for the transfer performance. Some previous studies demonstrate that, a delicate design of the balance proportions does influence the expected loss upperbound on target domain [22].

Another challenge comes from the possible category inconsistency among the sources. MSDA embraces more abundant resources than single-source DAs yet simultaneously, face the risk that easily causes negative transfers [24], due to multi-source category inconsistency. For instance, when categorical disalignment occurs among sources (as illustrated in Fig. 12.1c), different source domains may not precisely share their categories, thus, MSDA should take both category shift and domain shift into account.

To overcome the mentioned challenges, we propose *Cocktail Network* (DCTN), a transferable representation learning framework flexibly-deployed for solving MSDA across different transfer protocols. Suppose the classifier for each source domain is known. An ideal target example classifier in DCTN can be obtained by integrating all source predictions based on the corresponding source distribution weights. Therefore, besides of the feature extractor, DCTN also includes a (multi-source) category classifier to predict the classes from different sources, and a (multi-source) domain discriminator to produce multiple source-target-specific perplexity scores[1] as the approximation of source distribution weights.

Analogous to make cocktails, the multi-source category predictions are integrated with the perplexity scores to classify the target example, and thus the proposed method is dubbed by cocktail network (DCTN).

---

[1] Here perplexity scores are disconnected with the term used in natural language processing. In our chapter, they are completely determined by some relevant equations.

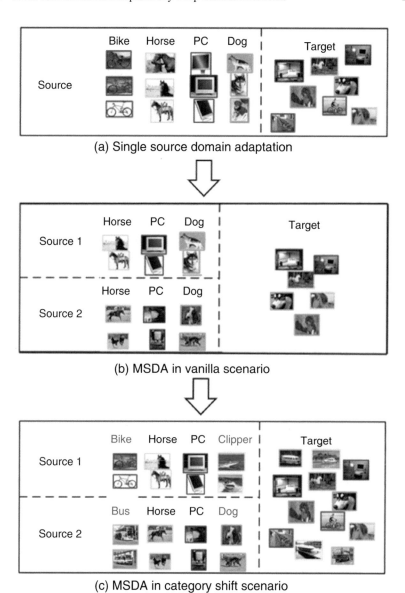

(a) Single source domain adaptation

(b) MSDA in vanilla scenario

(c) MSDA in category shift scenario

Fig. 12.1 The demonstrations of three domain adaptation scenarios. (**a**) Single source unsupervised domain adaptation (uDA) pressumes that all source examples are drawn from some underlying distribution under i.i.d. condition. (**b**) Vanilla MSDA scenarios consider multi-source and target examples that exactly share their categories. (**c**) MSDA in category shift pressume source data are collected from different source distributions with regards to different category sets

DCTN is learned by performing two alternating domain adaptation:

1. The multi-source domain discriminator is updated by using multi-way adversarial transfer learning to minimize the domain gaps between target and each source domains, so as to predict their corresponding multi-source perplexity scores;
2. The feature extractor and the category classifier are discriminatively fine-tuned with multi-source labeled and target pseudo-labeled data.

In this learning manner, the multi-way adversarial transfer learning implicitly reduces domain shifts between those sources. The discriminative adaptation helps to learn more classifiable features [28], and prevents the negative transfer effects caused by the mis-matching categories.

Delve deeper into our target data classification principle and we show that, this manner is coincidentally consistent with a well-known MSDA theory, i.e., *source distribution weighted combining rule* [22], where the target distribution is supposed to be the linear combination of multi-source distributions. The old-fashioned theory leads to a upperbound of expected classification error on the target domain. However, such multi-source weight combination involves a set of hyper-parameters precisely crafted by empirical experience, thus, is impractical under the deep domain adaptation background. Comparatively, DCTN employ a multi-way adversarial learning to automatically decide the balance weights among the sources.

We conduct some empirical evaluations for DCTN. They are conveyed on three DA benchmarks, i.e., Office-31, Image-CLEF and Digit-five that correspond to 2-to-1, 2-to-1 and 4-to-1 multi-source transfer learning cases, respectively. Under the MSDA vanilla protocol, DCTN performs quite competitive to some state-of-the-art deep DA baselines. Under the MSDA category shift protocol, transfers performed by DCTN are considered via varying a proportion of public classes of multiple sources. In this case, deep DA baselines almost suffer negative transfer and their results degrade while DCTN always keeps positive transfer performance. These experiments are consistent with a fact that, our DCTN is a superior framework for solving MSDA from diverse transfer learning aspects. This chapter is partially based on earlier work from [32].

12.2   Problem Setups

Two MSDA scenarios are to be specified in this section. The first is common in most existing studies, yet the second and third come from the recent advance in single source DA.

12.2.1 Vanilla MSDA Transfer Scenario

In terms of multi-source transfer, there are M different source distributions denoted by $\{P_j(x, y)\}_{j=1}^{M}$ with labeled source domain images $\{(\mathbf{X}_j, \mathbf{Y}_j)\}_{j=1}^{M}$ drawn from those distributions respectively, where $\mathbf{X}_j = \{x_{j,i}\}_{i=1}^{N_j}$ represents N_j images from source j in total and $\mathbf{Y}_j = \{y_{j,i}\}_{i=1}^{N_j}$ corresponds to their labels. Besides, there is a target distribution $P_t(x, y)$, where target image set $\mathbf{X}^{(t)} = \{x_i^{(t)}\}_{i=1}^{N^{(t)}}$ are drawn from yet without label observation $\mathbf{Y}^{(t)}$. Those $M+1$ domain sets have been treated as one training set ensemble, and the test set $(\mathbf{X}_{test}, \mathbf{Y}_{test}) = \{x_i^{test}, y_i^{text}\}_{i=1}^{N_{test}}$ are drawn from $P_t(x, y)$ to evaluate the model adaptation performance.

12.2.1.1 Category Shifts

Under the vanilla MSDA setting, samples from diverse sources share a same category set. In contrast to this convention, we introduce a new MSDA scenario where the categories from M sources are also different. Formally speaking, given a category set $\mathcal{C}_j = \bigcup_{i=1}^{|\mathbf{Y}_s|} \{y_{j,i}\}$ as a class set of $\mathbf{Y}_j$ for domain j, the relation between $\mathcal{C}_j$ and $\mathcal{C}_k$ has been generalized from $\mathcal{C}_j \cup \mathcal{C}_k = \mathcal{C}_j \cap \mathcal{C}_k$ to $\mathcal{C}_j \cap \mathcal{C}_k \subseteq \mathcal{C}_j \cup \mathcal{C}_k$, where $\mathcal{C}_j \cap \mathcal{C}_k$ denotes the public classes between sources j and k. Suppose data in a target domain are labeled by the union of categories in those sources $(\mathcal{C}_t = \bigcup_{j=1}^{M} \mathcal{C}_j)$, then we term $\mathcal{C}_j \cap \mathcal{C}_k \neq \mathcal{C}_j \cup \mathcal{C}_k$ as *category shift* in multiple source domains $\{(\mathbf{X}_j, \mathbf{Y}_j)\}_{j=1}^{M}$.

12.3 Deep Cocktail Networks (DCTNs)

MSDAs are challenging to tackle neither in vanilla nor the other multi-source transfer scenarios. The main difficulty comes from keeping balance between the sources that may affects the target data classification. Intuitively, if a source domain share more common characteristics with the target domain, their labeled data should be expected to deliver more reliable discriminative information than those drawn from other sources. Our solution, i.e., Deep CockTail Networks (DCTNs) is proposed by this motivation. The framework of DCTN has been concisely shown in Fig. 12.2, and we elaborate it as follow.

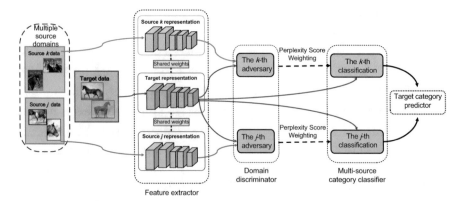

Fig. 12.2 An overview of the proposed CockTail Network (DCTN). Our framework receives multi-source instances with annotated ground truth and adapts to classify the target samples. Let's confine problem with only source j and k for simplicity. (1) The feature extractor maps target, source j and k into a common feature space. (2) The category classifier receives target feature and produces the j-th and k-th classifications based upon the categories in source j and k respectively. (3) The domain discriminator receives features from source j, k and target, then offers the k-th advesary between target and source k, as well as the j-th advesary between target and source j. The j-th and k-th advesary provide source j and k perplexity scores to weight the j-th and k-th classifications correspondingly. (4) The target classification operator integrates all weighted classification results then predicts the target class across category shifts. Best viewed in color

12.3.1 Framework

The framework of DCTN consists of four parts: three subnets, i.e., *feature extractor*, *(multi-source) domain discriminator*, *(multi-source) category classifier*; and a *target classification operator*, the rule combining the three subnets to classify target samples.

Feature extractor F, a deep convolution network is introduced as the backbone, mapping images from M sources and target into a common feature space. Since adversarial learning performs well in learning both domain-invariant features and each target-source-specific relations, we employ a domain-shared pipeline to obtain the optimal and scalable transfer mapping.

(Multi-source) domain discriminator D, a set of M source-specific discriminators $\{D_j\}_{j=1}^{M}$, is built for adversary. Given an image x, from the source j or the target domain, should first go through the feature extractor F, then the domain discriminator D receives the output features $F(x)$. Afterwards the source-specific discriminator D_j classifies whether $F(x)$ originates from the source j or the target. The data flows from each source will not trigger those discriminators from the other sources, while the data flows from each target instance $x^{(t)}$ makes the domain discriminator D yield M source-specific discriminative results $\{D_j(F(x^{(t)}))\}_{j=1}^{M}$ each of which corresponds to a source domain. They are leveraged to update the discriminator D, and also to supply the target-source perplexity scores $\{s(x^{(t)}; F, D_j)\}_{j=1}^{M}$ to the target classification operator

$$s(x^{(t)}; F, D_j) = -\log(1 - D_j(F(x^{(t)}))) \tag{12.1}$$

(Multi-source) category classifier C, a multi-output net composed by M source-specific category predictors $\{C_j\}_{j=1}^M$, is incorporated to do categorical classification. Each predictor is a softmax classifier configured by the category set in the corresponding source j. Given image x, the category classifier takes its mapping from feature extractor as input, then: (1) if the image is from source j, only the value from C_j get activated and provides the gradient for training; (2) if it is a target image $x^{(t)}$, all source-specific predictors provide M categorization results $\{C_j(F(x^{(t)}))\}_{j=1}^M$ to the target classification operator.

Target classification operator, the key to classify target examples, is led into the setting to assign the final label to each target sample. More specifically, for each target feature $F(x^{(t)})$, the target classification operator takes each source perplexity score $s(x^{(t)}; F, D_j)$ to re-weight the corresponding source-specific prediction $C_j(F(x^{(t)}))$, then accumulates the results to classify target example $x^{(t)}$. If the class $c \in \bigcup\limits_{j=1}^M \{C_j\}$ is considered, the confidence $x^{(t)}$ belongs to c presents as

$$C_t(c|x^{(t)}) := \sum_{c \in C_j} \frac{s(x^{(t)}; F, D_j)}{\sum\limits_{c \in C_k} s(x^{(t)}; F, D_k)} C_j(c|F(x^{(t)})) \tag{12.2}$$

where $C_j(c|F(x^{(t)}))$ is the softmax-based entry value of source j corresponding to class c. $x^{(t)}$ is categorized into the class with the highest confidence. The sum $\sum\limits_{c \in C_j}$ means only those sources with class c can join the perplexity score re-weighting. It's invented to incorporate both the vanilla and the *multi-source category shift* settings. Since the module independently estimates each class confidence, the variation in those shifted categories merely modifies the class combination in the target classification operator, but not the structures or the parameters in the three subnets. This design can be flexibly-deployed in vanilla and category shift scenarios.

12.3.2 Learning

DCTN admits an alternative adaptation pipeline. Briefly, after pre-training the feature extractor and classifier,[2] DCTN incorporates the aforementioned multi-way adversary to acquire a mutual mapping from all domains, then the feature extractor and the category classifier are fine-tuned with M-source labeled and target pseudo-labeled examples, attempting to endow the domain-invariant with more discriminative power. The two steps alternatively repeat until the epoch limit is reached.

---

[2] The pre-training process can be found in the original paper and the official code in.

12.3.2.1 Multi-Way Adversarial Adaptation

Multi-way adversarial adaptation is proposed to obtain domain-invariant features by proposing domain adaptation from each source to the target domain. It is described as a follow empirical loss objective:

$$\min_F \max_D V(F, D; \overline{C}) = \mathcal{L}_{adv}(F, D) + \mathcal{L}_{cls}(F, \overline{C}) \tag{12.3}$$

where

$$\mathcal{L}_{adv}(F, D) = \frac{1}{M} \sum_j^M \Big[\mathbb{E}_{x \sim \mathbf{X}_j}[\log D_j(F(x))] \tag{12.4}$$

$$+ \mathbb{E}_{x^{(t)} \sim \mathbf{X}_t}[\log(1 - D_j(F(x^{(t)})))] \Big]$$

where the first term denotes our multi-way adversarial mechanism; the second term indicates M cross-entropy losses for source-specific classification. The classifier C is fixed as $\overline{C}$ to provide stable gradient values.

It is worth noting that, the optimization on Eq. 12.3 works well for D but not F. Since the feature extractor learns to capture the mapping from the M sources to the target, the domain distributions become simultaneously changing in adversary, resulting an oscillation that may become detrimental to our feature extractor. Towards such concern, Tzeng et al.[29] mentioned when source and target feature mappings share their architectures, the domain confusion can be employed to replace the adversarial objective, which more stably learns the mapping F. Extending it to our scenario, we devise the multi-domain confusion loss:

$$\mathcal{L}_{adv}(F, D) = \frac{1}{M} \sum_j^M \Big[\mathbb{E}_{x \sim \mathbf{X}_j} \mathcal{L}_{cf}(x; F, D_j) \tag{12.5}$$

$$+ \mathbb{E}_{x \sim \mathbf{X}_t} \mathcal{L}_{cf}(x^{(t)}; F, D_j) \Big]$$

where

$$\mathcal{L}_{cf}(x; F, D_{s_j}) = \tag{12.6}$$

$$\frac{1}{2} \log D_j(F(x)) + \frac{1}{2} \log(1 - D_j(F(x)))$$

Hard Domain Batch Mining In the stochastic gradient-based optimization, the multi-way adversarial adaptation receives N samples from M sources to iteratively update the feature extractor F. However, the samples drawn from different sources do not always help to boost the adaptation performance. In fact, as the training proceeds, redundant source samples turn to draw back the previous adaptation results. To deal with this we design a simple yet effective hard domain batch mining technique to improve the training efficiency.

More specifically, suppose that in each iteration, DCTN randomly draws m target examples $\{x_i^{(t)}\}_{i=1}^m$ and m source examples for each source, i.e., $\{\{x_{1,i}\}_{i=1}^m, \cdots, \{x_{M,i}\}_{i=1}^m\}$. Hence there are totally $m(M+1)$ samples for training DCTN per iteration. We keep the discriminator training as described in Eq. 12.3. As for feature extractor training, we independently consider the adversary between the target and each source. Given this, a source-target discriminator loss $\sum_i^m - \log D_j(F(x_{j,i})) - \log[1 - D_j(F(x_i^{(t)}))]$ could be viewed as the "difficulty" degree to distinguish $x_i^{(t)}$ from m source-j samples. Therefore, if F performs worst to transform the target samples to confuse the source j^*, it results in $j^* = \arg\max_j\{\sum_i^m - \log D_j(F(x_{j,i})) - \log[1 - D_j(F(x_i^{(t)}))]\}_{j=1}^M$. Based upon the domain confusion loss, we use the source j^* and the target examples in the mini-batch to train the feature extractor F. This stochastic learning method is represented by the Algorithm 3.

12.3.2.2   Target Discriminative Adaptation

Sharing the spirit of existing work in adversarial DAs, the multi-way adversarial adaptation process does not consider the category variation during learning. Though DCTN has been able to learn domain-invariant features, this does not insure their classifiability in a target domain. David et al. [1] demonstrates a fact that, in order to accommodate a source classifier in the target, DA algorithms requires the category classifier working out across domains. It is more challenging in the context of MSDA since the domain-inconsistency exist not only between each source-target pair but also among multiple sources. Besides, in a variety of MSDA cases, their classifiers should account for the categorical misalignment across M sources and target.

Algorithm 3 Mini-batch learning via online hard domain batch mining

---

Input: Mini-batch $\{x_i^{(t)}, \{x_i^j, y_i^j\}_{j=1}^M\}_{i=1}^m$ sampled from X_t and $\{(X_j, Y_j)\}_{j=1}^M$ respectively; feature extractor F; domain discriminator D; category classifier $\overline{C}$.
Output: Updated F'.

1: Select source $j^* \in [M]$, where $j^* = \arg\max_j\{\sum_i^m - \log D_j(F(x_i^j)) - \log(1 -$
    $D_{s_j}(F(x_i^t)))\}_{j=1}^M$;
2: $\mathcal{L}_{adv}^{j^*} = \sum_i^m \mathcal{L}_{cf}(x_i^{j^*}; F, D_{j^*}) + \mathcal{L}_{cf}(x_i^t; F, D_{j^*})$
3: Replace $\mathcal{L}_{adv}$ in Eq. 12.3 with $\mathcal{L}_{adv}^{s_{j^*}}$, update F by Eq. 12.3.
4: **return** $F' = F$.

To achieve this universal target category predictor, we incorporate target examples to learn classifiable features with source data via discriminatively fine-tuning $\{C_j\}_{j=1}^{M}$ or $\{C_j^{(R)}\}_{j=1}^{M}$. We develop a switchable strategy to select and annotate target samples, after that, use them to train the feature extractor F and multi-source classifiers with multi-source labeled samples and pseudo-labeled target examples. In particular, we use the target category predictor obtained in the previous iter to annotate each target sample. The strategy selects suitable target examples $\mathbf{X}_t^{(p)}$ and assigns their pseudo labels $\{\mathbf{Y}_t^{(p)}\}$. The selection and annotation principle switches according to different scenarios. Specifically, the discriminative adaptation of DCTN presents as

$$
\min_{F, C} \mathcal{L}_{cls}(F, C) = \sum_{j}^{N} \mathbb{E}_{(x, y) \sim (\mathbf{X}_j, \mathbf{Y}_j)}[\mathcal{L}(C_j(F(x)), y)]
$$
$$
+ \mathbb{E}_{(x^{(t)}, \hat{y}^{(t)}) \sim (\mathbf{X}_t^{(p)}, \mathbf{Y}_t^{(p)})}[\sum_{\hat{y} \in \mathcal{C}_{\hat{s}}} \mathcal{L}(C_{\hat{s}}(F(x^{(t)})), \hat{y}^{(t)})]
$$
(12.7)

where the first and second terms denote the classification losses from multiple source images $\{\mathbf{X}_j, \mathbf{Y}_j\}_{j=1}^{M}$, and target images with pseudo labels $\{\mathbf{X}_t^{(p)}, \mathbf{Y}_t^{(p)}\}$ respectively. We apply the target classification operator to assign pseudo labels, and the samples with the confidence higher than a presented threshold γ will be selected into $\mathbf{X}_t^{(p)}$.

Since the target predictions come by integrating multi-source classification, there is no explicit learnable target classifier. As illustrated in the second term of Eq. 12.7, we apply the multi-source category classifier to back-propagate pseudo target classification errors. Concretely, given a target instance $x^{(t)}$ with pseudo-labeled class $\hat{y}$, we find those sources $\hat{s}$ including this class ($\hat{y} \in \mathcal{C}_{\hat{s}}$), then update our network via the sum of the multi-source classification losses, namely, $\sum_{\hat{y} \in \mathcal{C}_{\hat{s}}} \mathcal{L}(C_{\hat{s}}(F(x^{(t)})), \hat{y})$ in the second term.

Combing the aforementioned two processes, the adaptation algorithm of DCTN has been summarized in Algorithm 4.

12.3.3 Connection with the Source Distribution Weighted Combining Rule

Let $\{\mathcal{P}_j\}_{j=1}^{M}$ and $\mathcal{P}_t$ represent M source and target distributions.[3] Given an instance x, $\{\mathcal{P}_j(F(x))\}_{j=1}^{M}$ and $\mathcal{P}_t(F(x))$ denote the probabilities of x drawn from $\{\mathcal{P}_j\}_{j=1}^{M}$

---

[3]Since each sample x corresponds to an unique class y, $\{\mathcal{P}_j\}_{j=1}^{M}$ and $\mathcal{P}_t$ can be viewed as an equivalent embedding from $\{P_j(x, y)\}_{j=1}^{N}$ and $P_t(x, y)$ that we have discussed.

Algorithm 4 Learning algorithm for DCTN

Input: N source labeled datasets $\{\mathbf{X}_j, \mathbf{Y}_j\}_{j=1}^M$; target unlabeled dataset $\mathbf{X}_t$; initiated feature extractor F; multi-source category classifier C and domain discriminator D; confidence threshold γ; entropy threshold ζ; maximal adversarial iteration β.

Output: well-trained feature extractor F^*, domain discriminator D^* and multi-source category classifier C^*.

1: **Pre-train C and F**
2: **while** not converged **do**
3:      **Multi-way Adversarial Adaptation:**
4:      **for** $1{:}\beta$ **do**
5:          Sample mini-batch from $\{\mathbf{X}_j\}_{j=1}^M$ and $\mathbf{X}_t$;
6:          Update D by Eq. 12.3;
7:          Update F by Algorithm 3;
8:      **end for**
9:      **Target Discriminative Adaptation:**
10:      Estimate confidence for $\mathbf{X}_t$ by Eq. 12.2. Samples $\mathbf{X}_t^{(p)} \subset \mathbf{X}_t$ with confidence larger than γ get annotations $\mathbf{Y}_t^{(p)}$;
11:      Update F and C by Eq. 12.7.
12: **end while**
13: **return** $F^* = F$; $C^* = C$; $D^* = D$.

---

and $\mathcal{P}_t$, respectively. In the *source distribution weighted combining rule* [22], the target distribution can be defined as a mixture of the multi-source distributions with the coefficients by normalizing source distributions weighted by an implicit simplex $\triangle = \{\lambda : \lambda_j \geq 0 \,\&\, \sum_{j=1}^M \lambda_j = 1\}_{j=1}^M$, namely, $\mathcal{P}_t(F(\mathbf{x})) = \sum_{c \in \mathcal{C}_k}^M \lambda_k \mathcal{P}_k(F(\mathbf{x}))$. For simplicity in our discussion, we only consider the vanilla MSDA case so that $\mathcal{C}_k = \underset{j}{\overset{M}{\cup}} \mathcal{C}_j$, $\forall k \in [M]$. Under this target distribution combination assumption, the ideal target classifier $C_t(c|\mathbf{x}^{(t)})$ presents as the linear weighted combination of source classifiers $\{C_j(c|F(\mathbf{x}^{(t)}))\}_{j=1}^M$:

$$C_t(c|\mathbf{x}^{(t)}) = \sum_{j=1}^M \frac{\lambda_j \mathcal{P}_j(F(\mathbf{x}^{(t)}))}{\mathcal{P}_t(F(\mathbf{x}^{(t)}))} C_j(c|F(\mathbf{x}^{(t)})) \qquad (12.8)$$

Formally, we frame the formula into adaptation learning theory to provide more interpretations. In specific, $\mathcal{X}$ represents the input space (the feature space we consider); $f : \mathcal{X} \to \mathbb{R}$ denotes the target function to learn (correspond to the ground-truth); $h : \mathcal{X} \to \mathbb{R}$ denotes the hypotheses with respect to a specific underlying distribution (correspond to the classifier); $L : \mathbb{R} \times \mathbb{R} \to \mathbb{R}$ denote a convex loss function acting on a feature (correspond to classification error). Hence a learning problem is mostly formulated as

$$\min_{h \in \mathcal{H}} \mathcal{L}(\mathcal{P}, h, f) = \mathbb{E}_{\mathbf{x} \sim \mathcal{P}}[L(h(\mathbf{x}), f(\mathbf{x}))] \tag{12.9}$$

where according to the definition of $\{\mathcal{P}_j\}_{j=1}^M$ and $\mathcal{P}_t$, $\mathbf{x}$ denote the feature of x that $\mathbf{x} \sim \mathcal{P}(F(x))$. We would like to write it as $\mathbf{x} \sim \mathcal{P}(\mathbf{x})$. Suppose M source hypotheses $\{h_1, \cdots, h_M\}$ correspond to $\{\mathcal{P}_1, \cdots, \mathcal{P}_M\}$ and thus, for all $j \in [M]$, $\mathcal{L}(\mathcal{P}, h, f) \leq \epsilon$, $(\epsilon > 0)$, source distribution weight combining rule maintains a upperbound of expected loss on a target domain:

$$h_\lambda(x) = \sum_{j=1}^M \frac{\lambda_j \mathcal{P}_j(x)}{\mathcal{P}_t(x)} h_j(x) \ s.t. \ \mathcal{P}_t(x) = P_\lambda(x) = \sum_{j=1}^M \lambda_j \mathcal{P}_j(x), \ \forall \lambda \in \Delta \tag{12.10}$$

where given a target distribution $\mathcal{P}_t$ as a mixture P_λ of multiple source distributions $\{\mathcal{P}_j\}_{j=1}^M$ w.r.t.,λ, the expected loss of its mixture hypothesis h_λ is at most ϵ w.r.t. any target function f, i.e., $\mathcal{L}(P_\lambda, h_\lambda, f) \leq \epsilon$.

The mixture hypothesis h_λ easily corresponds to $C_t(c|x^{(t)})$ in Eq. 12.8. The theorem demonstrates that, if we are able to find ideal hyper-parameters $\lambda \in \Delta$ so that the target distribution can be represented as a mixture of multiple source distributions in Eq. 12.10, the target classifier is certified to keep an upperbound of target distribution without training with target labeled data.

Revisit our DCTN target classification principle in Eq. 12.2. It is interesting that our DCTN also leverages source-target relationships to aggregate the multi-source classification outcomes, a similar motivation to decide a target classifier. But compared with Eq. 12.8 in details, the weights chosen by DCTN is totally according to the outcomes from adversarial learning. Obviously, it would be very convenient for deep adversarial domain adaptation. Specifically, we first translate Eq. 12.2 into our analysis background, then it turns to

$$h_t(\mathbf{x}) = \sum_{j=1}^M \frac{-\log(1 - D_j^*(\mathbf{x}))}{\sum_{k=1}^M -\log(1 - D_k^*(\mathbf{x}))} h_j(\mathbf{x}) \tag{12.11}$$

where $\forall j \in [M]$, D_j^* denotes the optimal domain discriminator of the adversary between source j and target. Replace the M source-specific hypotheses by $\{C_j(c|F(x^{(t)}))\}_{j=1}^M$, Eq. 12.11 exactly degenerates into Eq. 12.2.

12.4   Experiments

In the context of MSDA for visual classification, we evaluate the accuracy of the predictions from the target classification operator in all experiments. Two transfer scenarios of MSDA, e.g., the vanilla MSDA setting, category shift have been considered. Our DCTN models are all implemented under the PyTorch[4] platform.

---

[4]http://pytorch.org/.

12.4.1   Experimental Setup

Three UDA benchmarks, e.g., *Office-31* [27], *ImageCLEF-DA*[5] and *Digits-five* are used for the MSDA evaluations. *Office-31* is an object recognition benchmark with 31 categories and 4652 images distributed across three visual data subsets **A** (*Amazon*), **D** (*DSLR*), **W** (*Webcam*). *ImageCLEF-DA* derives from domain adaptation challenge, and is composed of 12 object categories (aeroplane, bike, bird, boat, bottle, bus, car, dog, horse, monitor, motorbike, and people) shared in the three famous real-world datasets, **I** (*ImageNet ILSVRC 2012*), **P** (*Pascal VOC 2012*), **C** (*Caltech-256*). Each category includes 50 images. *Digits-five* includes five digit image sets drawn from public datasets **mt** (*MNIST*) [16], **mm** (*MNIST-M*) [6], **sv**(*SVHN*) [23], **up** (*USPS*) and **sy** (*Synthetic Digits*) [6]. Towards the images in *MNIST, MNIST-M, SVHN* and *Synthetic Digits* respectively. For each domain, we draw 25,000 for training and 9000 for testing. Since there are only 9298 images in *USPS*, we choose the entire benchmark as our domain. For real-world images, we follow the test routine in the previous works [18, 19] for fair comparisons; For digit dataset, the evaluation is performed on the test set in the target domain.

To perform the recognitions in Office-31 and ImageCLEF-DA, existing deep DA approaches [18, 19] routinely employ Alexnet [15] as their backbones. For a fair comparison, we choose a DCTN architecture derived from the Alexnet: The extractor F is designed as a five-layer fully-convolutional network with three max-pooling operators, and the (multi-source) category classifier C is a three-layer fully-connected multi-task network. They are stacked into an architecture exactly corresponding to the Alexnet pipeline. For our multi-way adversarial learning, we choose a CNN with a two-head classifier as domain discriminator D. Digits perceive less visual characteristic compared with real-world images. Hence we choose a more lightweight structures as our feature extractor F, category classifier C and domain discriminator D for digit recognition.

The implementation details can be found in the original paper [32].

12.4.2   Evaluations in the Vanilla Setting

Baselines The existing works of MSDA almost lack comprehensive evaluations on real-world visual recognition benchmarks. In our experiment, we introduce two baselines on shallow architecture: sparse FRAME (**sFRAME**) [30] and **SGF** [10] as the multi-source approaches in the *Office-31* experiment. Moreover, we also compare DCTN with single-source visual UDA methods including the conventional, e.g., Transfer Component Analysis (**TCA**) [25] and Geodesic Flow Kernel (**GFK**) [9], as well as state-of-the-art deep approaches: Deep Domain Confusion (**DDC**)

---

[5]http://imageclef.org/2014/adaptation.

[12], Deep Reconstruction-classification Networks (**DRCN**) [8], Reversed Gradient (**RevGrad**) [5], Domain Adaptation Network (**DAN**) [18], and Residual Transfer Network (**RTN**) [19] and Joint Adaptation Network (**JAN**) [20]. Since those methods are primitively invented for single-source setup, we use a *Source combine* standard for evaluating them in MSDA: all source domains are combined into a traditional single-source vs. target setting. The standard testify whether the multi-source is valuable to exploit. Additionally, as baselines in the *Source combine* and multi-source standards, we incorporate all images from sources to train backbone-based multi-source classifiers and apply them to classify target images. They are termed *Source only*, confirming whether the evaluated multi-source transfer learning fails or not.

Object Recognition We report results in all transfer cases and compare DCTN with the baselines. Tables 12.1 and 12.2 show that DCTN yields the best in the *Office-31* transfer tasks **A,W→D** and **A,D→W**, perform impressively in **D,W→ A**. Supported by deep representations, DCTN outperforms conventional MSDA baselines by a large margin. Besides, when the single source deep DA baselines, i.e., DAN, RTN, JAN and RevGred, deem the combined sources as a single domain, their source-combine performances are suppressed by the best single source baselines. In a comparison, DCTN basically outperforms all the baselines. It reveals that, if we treat MSDA as a single source DA problem, the performance gain cannot be fully excavated. But through the data transfer by DCTN, the potential power of multiple sources are successfully leveraged to boost the adaptation efficacy. In a comparison, the source-combine baselines in *ImageCLEF-DA* show more superior than their single source versions yet remains inferior to our DCTN. It validates that, no matter if domain size is equal or not, DCTN is still able to learn more transferable and discriminative features than the other baselines, from multi-source transfer for natural image domains.

Digit Recognition Distinct from the real-world visual recognition benchmarks, Digit-five consists of five domain datasets in total and is specified for multi-domain learning, therefore, we investigate 4-to-1 transfer learning results on DCTN by considering the domain shifts: **mm, mt, sy, up → sv** and **mt, sv, sy, up → mm**, and provide the performance on the average. We compare DCTN with RevGred, DAN and their source-combine transfer variants.

The results have been shown in Table 12.3. First of all, it is apparent that accuracies of single source DA approaches fall behind their source-combine. It implies that as M increases, multiple sources provide more edges to boost transfer performance gains than those solely involved with a single source domain. However, we observe that these source-combine typically perform worse than their source-only except for **mt, sv, sy, up → mm**. In other words, despite of potential benefits multiple sources bring about, single source deep DA approaches conventionally suffer negative transfer, therefore, are difficult to take advantage of the multi-source information into the model. In comparison, DCTN consistently shows positive transfer performances compared with the source only, and no matter of source-

Table 12.1 Classification accuracy (%) on Office-31 benchmark for MSDA in the vanilla setting

Standards	Models	W→D	A→D	A→W	D→W	D→A	W→A	Avg
Single source	TCA	95.2	60.8	61.0	93.2	51.6	50.9	68.8
	GFK	95.0	60.6	60.4	95.6	52.4	48.1	68.7
	DDC	98.5	64.4	61.8	95.0	52.1	52.2	70.7
	DRCN	99.0	66.8	68.7	96.4	**56.0**	54.9	73.6
	WMMD	98.7	64.5	66.8	95.9	53.8	52.7	72.1
	RevGrad	99.2	72.3	73.0	96.4	53.4	51.2	74.3
	DAN	99.0	67.0	68.5	96.0	54.0	53.1	72.9
	RTN	**99.6**	71.0	73.3	96.8	50.5	51.0	73.7
	JAN	99.5	71.8	74.9	96.6	**58.3**	55.0	76.0
		A,W→D		A,D→W		D,W→A		
Source combine	Source only	98.1		93.2		50.2		80.5
	RevGred	98.8		95.2		53.4		82.5
	DAN	98.8		95.2		53.4		82.5
	JAN	96.0		94.0		57.2		82.4
Multi-source	Source only	98.2		92.7		51.6		80.8
	RDALR	31.2		36.9		20.9		29.7
	sFRAME	54.5		52.2		32.1		46.3
	SGF	39.0		52.0		28.0		39.7
	DCTN (ours)	**100.0**		**96.9**		55.4		**84.1**

(The bold-value indicates the state-of-the-art performances.)

Table 12.2 Classification accuracy (%) on ImageCLEF-DA benchmark for MSDA in the vanilla setting

Standards	Models	I→P	C→P	I→C	P→C	P→I	C→I	Avg
Single source	RevGrad	66.5	63.5	89.0	88.7	81.8	79.8	78.2
	DAN	67.3	61.6	87.7	88.4	80.5	76.0	76.9
	RTN	67.4	63.0	89.5	90.1	82.3	78.0	78.4
	JAN	67.2	63.5	**91.3**	91.0	82.8	80.0	79.3
		I,C→P		I,P→C		P,C→I		
Source combine	Source only	68.3		88.0		81.2		79.2
	DAN	**68.8**		88.8		81.3		79.6
	JAN	**68.7**		89.4		82.6		80.2
Multi-source	Source only	68.5		89.3		81.3		79.7
	DCTN (ours)	**69.6**		**91.0**		**83.3**		**81.3**

(The bold-value indicates the state-of-the-art performances.)

combine and multi-source ensemble, DCTN always outperforms the other baselines. In Table 12.3, the mean accuracy of our DCTN exceeds the second best by 3.6%.

Feature Visualization In the experiment of adaptation task **A,D → W** in Office-31, we visualize the DCTN activations (the features ahead of classification) before and after adaptation. For simplicity, both the source domains A, D have been

Table 12.3 Classification accuracy (%) on Digits-five dataset for MSDA in the vanilla setting

Standards	Models	mm → sv	mt → sv	sy→ sv	up → sv	mt → mm	sv → mm	sy → mm	up → mm	Avg
Single source	Source only	45.3	46.4	67.4	29.7	58.0	49.6	54.8	43.7	49.4
	RevGred	45.3	46.4	67.4	29.7	58.0	49.6	54.8	43.7	49.4
	DAN	43.2	42.2	67.1	38.5	53.5	51.8	58.8	40.5	49.5
		mm, mt, sy, up → sv				mt, sv, sy, up → mm				
Source combine	Source only	72.2				64.1				68.2
	RevGred	62.6				62.9				62.8
	DAN	71.0				66.6				68.8
Multi-source	Source only (e)	64.6				60.7				62.7
	RevGred (e)	62.6				62.9				62.8
	DAN (e)	62.6				62.9				62.8
	DCTN (ours)	**78.7**				**71.6**				**75.2**

(e) indicates the ensemble version of the original baselines

Fig. 12.3 The t-SNE [21] visualization of A,D→W. Green, black and red represent domains A, D and W respectively. Different markers denote different categories, e.g., bookcase, calculator, monitor, printer, ruler. Best viewed in color

separated to emphasize the contrast of target. As we can see in Fig. 12.3, compared with the activations given by the source only, both of the activations from **A → W** and **D → W** have shown good adaptation patterns. It implies that DCTN can successfully learn transferable features with multiple sources. Besides, the target activations become more clear to categorize, which suggests that the features learned by DCTN attains desirable discriminative property. Finally, even if the multi-source transfer has been composed of hard transfer task (**A → W**), DCTN still enable target data adaptation without the degradation in the performance of **D → W**.

12.4.3   Evaluations in the Category Shift Setting

In this subsection, we turn to evaluate DCTN in the category shift scenario, where the multiple sources are not promised to share their categories. We compare our DCTN with Source-only (source-combine) and the other two state-of-the-art baselines, i.e., DAN (source-combine), RevGred (source-combine), based on their accuracies after the transfers happen in the four MSDA transfer cases: A,D→W in Office-31 and I,P→C in ImageCELF-DA.

How to Evaluate? Since source-category-shift is newly proposed in MSDA scenario, benchmarks should be amended to evaluate DA algorithms in this scenario. Specifically, suppose that M sources involve C categories and $C_p \leq C$ indicates the number of their public classes. Due to $M = 2$, we consider the alphabetical order of the C classes and take the first $\frac{C-C_p}{2}$ and last $\frac{C-C_p}{2}$ classes as the source-specific private classes, then the rest proportion $\frac{C_p}{C}$ denotes the public classes. To unveil the comprehensive the baselines in this scenario, we evaluate them by specifying the public-class proportions $\frac{C_p}{C}$ in $\{0, 0.3, 0.5, 0.7, 1\}$, respectively.

We elaborate three metrics to reflect the adaptation capability of baselines from different perspectives. First, classification accuracy is to evaluate whether the baseline helps the classifier get rid of domain/category shifts. Besides, since practitioners are more interested in how much performance drops when source-category shift exists, we employ a relative measure termed *degraded accuracy*, simply calculated as follows:

$$DA(\frac{C_p}{C}) = Acc(\frac{C'_p}{C} = 1) - Acc(\frac{C_p}{C}) \qquad (12.12)$$

where $Acc(\frac{C_p}{C})$ denotes the model's accuracy when the public-class proportion is $\frac{C_p}{C}$, then $Acc(\frac{C'_p}{C} = 1)$ means the accuracy of the model trained in vanilla MSDA. The formula showcases the performance drop caused by inconsistent categories of sources. The lower value means the algorithms less affected by this negative effect, performing more robust in this scenario. Finally, we employ *transfer gain* to further confirm the availability of transfer learning. Transfer gain is calculated through subtracting the baseline's accuracy with the accuracy of Source only. Higher is better. A positive value undoubtedly means that the transfer is available while a negative value means the DA approach return to enlarge the domain shift.

Results The evaluation covers the two transfer cases are illustrated in Figs. 12.4, 12.5, and 12.6, which denote the performances based on mean accuracy, degraded accuracy and transfer gain, respectively. DCTN always outperforms the other baselines in different proportions of public classes and transfer cases. Generally, the improvement becomes larger as the sources contain more public classes. Consider the relative enhancement measured by DA (Fig. 12.5), and then we discover that both Source only and DCTN behave neck and neck in ImageCELF-DA. In Office-31, Source-only even obtain lower DA values than DCTN. Note that, it does not imply Source-only superior than our DCTN. In particular, compared with DA algorithms, i.e., DAN, RevGred and DCTN, Source-only undergoes fully-supervised learning, therefore, is free of the risk caused by category misalignment. To some extent, it should be treated as a sort of consecutive strategy preferring the safety of supervised training rather than adapting to a domain without labeled data. Retrospect to the absolute performance of Source-only in Fig. 12.4 and it is obvious that, Source-only are almost inferior to all DA approaches.

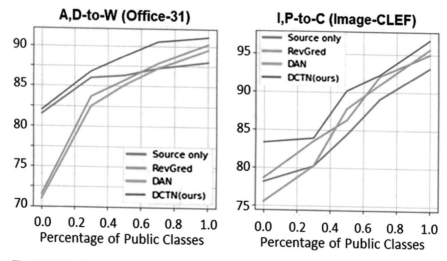

Fig. 12.4 The absolute performances based upon the mean accuracies (%) of Source only, RevGread, DAN and DCTN on Office-31 and Image-CLEF under the MSDA category shift scenario. The curves denote their accuracies changing as the public classes across multiple sources increase. Higher is better

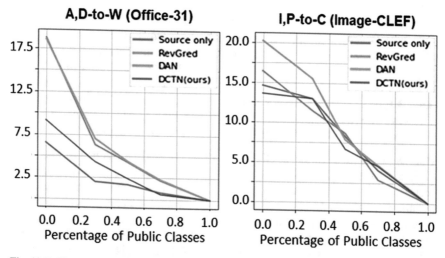

Fig. 12.5 The relative performance (degraded accuracies, the accuracy under vanilla scenario minus the accuracy under category shift) of Source only, RevGread, DAN and DCTN on Office-31 and Image-CLEF under MSDA category shift scenario. The curves denote how much their accuracies drop as the public classes across multiple sources increase. Lower is better

Beside of boosting transfer performance, another outstanding merit of DCTN is the strong resistance against the potential negative transfer influences. As demonstrated in Fig. 12.6, DCTN mostly outperforms the other baselines and more importantly, remain positive values in all transfer cases. More specifically, in

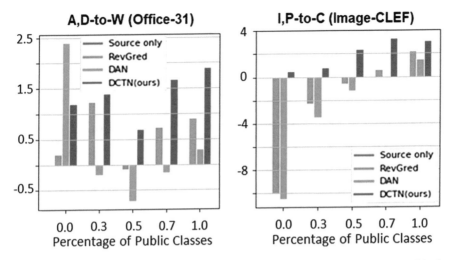

Fig. 12.6 The transfer gains (the accuracy of the baseline minus the accuracy of source only) of Source only, RevGread, DAN and DCTN on Office-31 and Image-CLEF under MSDA category shift scenario. The negative value means the negative transfer, which causes even heavier model damage than those without domain adaptation. Higher is better

Office-31, DAN show quite stunning transfer performance in A,D→W with 0% public classes, while its transfer gain is quite unstable as the public class alters. RevGred performs more stable and better than DAN in general, whereas both of them inevitably suffer from negative transfer and are wholly suppressed by our DCTN. In ImageCLEF-DA, DAN and RevGred hardly provide promising transfer performances. Ironically, when the number of public classes is small, their transfers even result in more model damages than the domain shift. In comparison, our DCTN always benefits the model through transfer learning.

12.5 Conclusion

In this chapter, we have explored the unsupervised domain adaptation with multiple source domains. It is an old topic in domain-adaptation area, whereas recent studies seldom work out due to the intrinsic difficulty of this adaptation problem preventing the success in deep feature learning. We raise a new MSDA framework termed *deep cocktail network*, to learn transferable and discriminative deep features from multiple sources. The approach can be applied to the ordinary MSDA setting and category shift, where classes from different sources are non-consistent. Empirical results show the impressive performances of DCTN across different transfer scenarios.

Acknowledgments We would like to thank the other authors, i.e., Ruijia Xu, Wangmeng Zuo and Junjie Yan, for their contribution to the original paper.

References

1. Ben-David, S., Blitzer, J., Crammer, K., Kulesza, A., Pereira, F., Vaughan, J.W.: A theory of learning from different domains. Mach. Learn. **79**(1), 151–175 (2010)
2. Bousmalis, K., Silberman, N., Dohan, D., Erhan, D., Krishnan, D.: Unsupervised pixel-level domain adaptation with generative adversarial networks (2016). Preprint. arXiv:1612.05424
3. Deng, J., Dong, W., Socher, R., Li, L.J., Li, K., Fei-Fei, L.: Imagenet: a large-scale hierarchical image database. In: IEEE Conference on Computer Vision and Pattern Recognition, 2009. CVPR 2009, pp. 248–255. IEEE, Piscataway (2009)
4. Duan, L., Xu, D., Tsang, I.W.H.: Domain adaptation from multiple sources: a domain-dependent regularization approach. IEEE Trans. Neural Netw. Learn. Syst. **23**(3), 504–518 (2012)
5. Ganin, Y., Lempitsky, V.: Unsupervised domain adaptation by backpropagation. In: International Conference on Machine Learning, pp. 1180–1189 (2015)
6. Ganin, Y., Ustinova, E., Ajakan, H., Germain, P., Larochelle, H., Laviolette, F., Marchand, M., Lempitsky, V.: Domain-adversarial training of neural networks. J. Mach. Learn. Res. **17**(1), 2096–2030 (2016)
7. Gebru, T., Hoffman, J., Fei-Fei, L.: Fine-grained recognition in the wild: a multi-task domain adaptation approach (2017). Preprint. arXiv:1709.02476
8. Ghifary, M., Kleijn, W.B., Zhang, M., Balduzzi, D., Li, W.: Deep reconstruction-classification networks for unsupervised domain adaptation. In: European Conference on Computer Vision, pp. 597–613. Springer, Cham (2016)
9. Gong, B., Shi, Y., Sha, F., Grauman, K.: Geodesic flow kernel for unsupervised domain adaptation. In: 2012 IEEE Conference on Computer Vision and Pattern Recognition (CVPR), pp. 2066–2073. IEEE, Piscataway (2012)
10. Gopalan, R., Li, R., Chellappa, R.: Domain adaptation for object recognition: An unsupervised approach. In: 2011 IEEE International Conference on Computer Vision (ICCV), pp. 999–1006. IEEE, Piscataway (2011)
11. Gretton, A., Smola, A., Huang, J., Schmittfull, M., Borgwardt, K., Schölkopf, B.: Covariate shift by kernel mean matching. Dataset Shift Mach. Learn. **3**(4), 5 (2009)
12. Hoffman, J., Tzeng, E., Darrell, T., Saenko, K.: Simultaneous deep transfer across domains and tasks. In: Domain Adaptation in Computer Vision Applications, pp. 173–187. Springer, Cham (2017)
13. Jhuo, I.H., Liu, D., Lee, D., Chang, S.F.: Robust visual domain adaptation with low-rank reconstruction. In: 2012 IEEE Conference on Computer Vision and Pattern Recognition (CVPR), pp. 2168–2175. IEEE, Piscataway (2012)
14. Krishna, R., Zhu, Y., Groth, O., Johnson, J., Hata, K., Kravitz, J., Chen, S., Kalantidis, Y., Li, L.J., Shamma, D.A.: Visual genome: connecting language and vision using crowdsourced dense image annotations. Int. J. Comput. Vis. **123**(1), 32–73 (2017)
15. Krizhevsky, A., Sutskever, I., Hinton, G.E.: Imagenet classification with deep convolutional neural networks. In: Advances in Neural Information Processing Systems, pp. 1097–1105 (2012)
16. LeCun, Y., Bottou, L., Bengio, Y., Haffner, P.: Gradient-based learning applied to document recognition. Proc. IEEE **86**(11), 2278–2324 (1998)
17. Long, J., Shelhamer, E., Darrell, T.: Fully convolutional networks for semantic segmentation. In: Proceedings of the IEEE Conference on Computer Vision and Pattern Recognition, pp. 3431–3440 (2015)

18. Long, M., Cao, Y., Wang, J., Jordan, M.: Learning transferable features with deep adaptation networks. In: International Conference on Machine Learning, pp. 97–105 (2015)
19. Long, M., Zhu, H., Wang, J., Jordan, M.I.: Unsupervised domain adaptation with residual transfer networks. In: Advances in Neural Information Processing Systems, pp. 136–144 (2016)
20. Long, M., Zhu, H., Wang, J., Jordan, M.I.: Deep transfer learning with joint adaptation networks. In: Proceedings of the 34th International Conference on Machine Learning, vol. 70, pp. 2208–2217 (2017). JMLR.org
21. Maaten, L.v.d., Hinton, G.: Visualizing data using t-SNE. J. Mach. Learn. Res. **9**(Nov), 2579–2605 (2008)
22. Mansour, Y., Mohri, M., Rostamizadeh, A.: Domain adaptation with multiple sources. In: Advances in Neural Information Processing Systems, pp. 1041–1048 (2009)
23. Netzer, Y., Wang, T., Coates, A., Bissacco, A., Wu, B., Ng, A.Y.: Reading digits in natural images with unsupervised feature learning. In: NIPS Workshop on Deep Learning and Unsupervised Feature Learning (2011)
24. Pan, S.J., Yang, Q.: A survey on transfer learning. IEEE Trans. Knowl. Data Eng. **22**(10), 1345–1359 (2010)
25. Pan, S.J., Tsang, I.W., Kwok, J.T., Yang, Q.: Domain adaptation via transfer component analysis. IEEE Trans. Neural Netw. **22**(2), 199–210 (2011)
26. Ren, S., He, K., Girshick, R., Sun, J.: Faster r-cnn: Towards real-time object detection with region proposal networks. In: Advances in Neural Information Processing Systems, pp. 91–99 (2015)
27. Saenko, K., Kulis, B., Fritz, M., Darrell, T.: Adapting visual category models to new domains. In: Computer Vision–ECCV 2010, pp. 213–226 (2010)
28. Saito, K., Ushiku, Y., Harada, T.: Asymmetric tri-training for unsupervised domain adaptation (2017). Preprint. arXiv:1702.08400
29. Tzeng, E., Hoffman, J., Saenko, K., Darrell, T.: Adversarial discriminative domain adaptation (2017). Preprint. arXiv:1702.05464
30. Xie, J., Hu, W., Zhu, S.C., Wu, Y.N.: Learning sparse frame models for natural image patterns. Int. J. Comput. Vis. **114**(2–3), 91–112 (2015)
31. Xu, K., Ba, J., Kiros, R., Cho, K., Courville, A., Salakhudinov, R., Zemel, R., Bengio, Y.: Show, attend and tell: neural image caption generation with visual attention. In: International Conference on Machine Learning, pp. 2048–2057 (2015)
32. Xu, R., Chen, Z., Zuo, W., Yan, J., Lin, L.: Deep cocktail network: multi-source unsupervised domain adaptation with category shift. In: Proceedings of the IEEE Conference on Computer Vision and Pattern Recognition, pp. 3964–3973 (2018)
33. Yang, J., Yan, R., Hauptmann, A.G.: Cross-domain video concept detection using adaptive svms. In: Proceedings of the 15th ACM International Conference on Multimedia, pp. 188–197. ACM, New York (2007)

Chapter 13
Zero-Shot Task Transfer

Arghya Pal and Vineeth N. Balasubramanian

13.1 Introduction

Curated labeled data is the key driving force of today's computer vision, machine learning and deep learning applications. State-of-the-art deep networks utilize a huge amount of curated labeled data and learn non-linear function approximations by optimizing hundreds of thousands of parameters. However, acquisition of curated labeled data for several real-world vision tasks invites long hours of human labor, expensive sensor-based data collection, intensive expertise in data curation and labeling—which collectively make labeled data collection a costly and time consuming assignment [42, 53]. As a result, many vision tasks are regarded as expensive and practitioners either work with noisy labeled data or low amount of curated labeled data, often leading to models with poor performance. In this chapter, we address this problem by proposing an alternate approach, viz., learning a meta-learner that can obtain model parameters for tasks without any labeled data. We refer to our proposed method as *zero-shot task transfer*, a novel extension to the existing zero-shot learning setting from class-level recognition to a vision task-level regression.

Classified as "self-produced movement", cognitive science has shown results [21, 47] where human subjects (especially human children) and few other mammals adapt to an entirely novel task (say, depth estimation) from their cognitive understanding and association of old learned tasks (self-motion, shoulder movements) to the novel task—without receiving an explicit supervision to the novel task. Such a

A. Pal (✉) · V. N. Balasubramanian
Indian Institute of Technology, Hyderabad, India
e-mail: cs15resch11001@iith.ac.in

© Springer Nature Switzerland AG 2020
H. Venkateswara, S. Panchanathan (eds.), *Domain Adaptation in Computer Vision with Deep Learning*, https://doi.org/10.1007/978-3-030-45529-3_13

finding supports the claim that a generalization to a novel cognitive task is possible by operating on the knowledge of some foundational tasks and their underlying relation with the novel task and the foundational tasks. Based upon this premise, we propose a meta-learning algorithm that will regress novel zero-shot task (i.e. tasks for which no ground truth is available) parameters by exploiting the knowledge of a set of already known tasks. Specifically, our proposed meta-learning algorithm will utilize the encoder-decoder model parameters of known tasks (i.e. tasks with ground truth) and learn to regress the encoder-decoder model parameters of zero-shot tasks. Formally speaking, given the knowledge of the encoder-decoder model parameters of m known tasks, $\{\tau^1, \tau^2, \cdots, \tau^m\}$, a meta-learning network $\mathcal{F}(.)$ regresses the encoder-decoder model parameters of a zero-shot task, $\tau^{(m+1)}$.

However, building a meta-learner *only* from the knowledge of known tasks is not plausible, since the meta-learner is free to converge to any undesired point on the meta-manifold (the manifold of model parameters). We hence introduce a pair-wise task correlation between source tasks and zero-shot tasks as an additional input to the meta-learning network $\mathcal{F}(.)$. In our framework, the pair-wise task correlation acts as a conditional input that assists the meta-learning network $\mathcal{F}(.)$ to converge to the desired encoder-decoder model parameters. There could be a variety of plausible opinions on how to obtain pair-wise task correlations. The present work utilizes the "wisdom of crowd" to acquire these pair-wise task correlations, as an example of achieving the proposed objective. Encoding the wisdom-of-crowd as a part of the learning process has found its foothold in several vision [38] and non-vision [42, 49] machine learning applications. The opinions of the crowd are integrated using standard high-fidelity vote aggregation methods such as the Dawid-Skene vote aggregation algorithm [14]. Towards the end of the discussion, we show that our proposed methodology is not limited to pair-wise task correlation acquired by aggregating the "wisdom of crowd" on task correlation, but can encapsulate any other task correlation acquired from other source, such as from [54].

Exploiting the task correlation as a part of learning the meta-learner might appear similar to the work of Taskonomy [54], but our proposed method and end objectives are different in many ways: (1) Taskonomy is a known-to-known task transfer approach and as an additional point, the work allows transfer of knowledge from multiple known source tasks to a single known target task, while, our method regresses the encoder-decoder parameters of a zero-shot task from the knowledge of known task parameters and the pair-wise task correlations; (2) In order to transfer to a novel task, Taskonomy requires a considerable amount of target labeled data; on the other hand, our framework regresses a novel task without any target labeled data; (3) The main objective of the Taskonomy work is to build a task graph based on the optimum transfer between one known task to another known task; while our approach leverages pair-wise task relation to regress zero-shot task model parameters; (4) Our method can regress multiple zero-shot tasks. This chapter is partially based on earlier work from [39].

Our contributions can be summarized as follows:

1. A meta-learning algorithm to regress high-performing model parameters of vision tasks with no labeled data.
2. The proposed meta-learner can regress model parameters of multiple zero-shot tasks. With a smaller set of known task parameters and task relations, our methodology infers high-performing model parameters of multiple zero-shot tasks, closer to (or in some cases, better than) the state-of-the-art model performance (learned with ground truth).
3. We also show results where the regressed model parameters of zero-shot tasks used in a transfer learning setting.

13.2 Related Work

Transfer Learning The methods of transfer learning assume relatedness of tasks and provide principled ways to discover and then reuse supervision from one task to another task. From the survey of [40], we can conclude that there are major four different ways to accomplish transfer learning: (1) *Instance-based transfer learning* deals with re-weighting of labeled data of the source domain and re-use them to a target domain. TrAdaBoost [12] or the active learning based method [22] are the names of some celebrated methods of instance-based transfer learning approach. Different from this, (2) *Relational network-based transfer* methods try to map a relation network such as a Markov Logic Network (MLN) [34], to help transfer from source to the target domain. On the other hand, (3) *Parameter sharing-based transfer* methods discover shared parameters or priors [9, 25] as a way to transfer knowledge. Lastly, (4) *Feature similarity-based transfer* methods discover transferable features as a way to transfer knowledge. Experiments on Convolutional Neural Network (CNN) features [52] have made it clear that the initial layers can be reused across tasks, as their filters show similarity. Recently, Zamir et al. [54] build a task graph of 26 vision tasks based on the optimum transfer from one source task to target task. However, unlike our method, none of the above methods can adapt to a novel task without any access to the ground truth of the novel task.

Multi-task Learning In the paradigm of multi-task learning, a set of related tasks are learned simultaneously with a view of task generalization. Rather than independently learning tasks, a common representation that is relevant to all tasks improves the final performance and generalization. However, one caveat on task sharing is the effect of negative transfer [25], where unrelated tasks can end up affecting each other and resulting to a degradation in performance. To mitigate this, several methods have been proposed to exploit the notion of task grouping. Some efforts assume a prior [9, 13] and then iterate over it, while other methods assume task parameters lie on a low-dimensional manifold [1], or a common subspace [25]. Distributed multi-task learning [30] focuses on learning multiple tasks distributed over a shared network. The main binding factor of all these methods is the access to ground truths of all tasks. Unlike our method, they cannot adapt to a zero-shot task with no labeled samples.

Domain Adaptation Primarily, domain adaptation methods deal with the problem of transferring knowledge from a data rich domain to a data scarce domain [11, 31]. Domain adaptation methods are concerned with learning domain-invariant features with a view of domain alignment. Such adaptation techniques mainly differ by the notion of domain alignment. For example, [18] performs domain adaptation by aligning the mid-level features of a CNN, while a method such as [18] uses an autoencoder-based technique to transfer related knowledge from the source to the target domain. Other methods include a clustering-based approach [46], or more recently, by using generative adversarial networks [29]. Recent efforts in unsupervised domain adaptation [8, 45] exploit the source and target domain discrepancy for a successful adaptation. However, a considerable amount of labeled data from both domains is still unavoidable.

Meta-learning Meta-learning is best described in literature as the process of *learning to learn* [32]. Black-box adaptation based methods use a neural network approach to infer parameters [20, 35]. Other efforts assume that task parameters lie on a low-dimensional subspace [3], share a common probabilistic prior [26], etc. On the other hand, [17, 23] are examples of optimization-based methods. Recent meta-learning approaches consider all task parameters as input signals to learn a meta manifold that helps few-shot learning [36, 48], transfer learning [44] and domain adaptation [18]. Unlike our method, all of the aforementioned methods assume an availability of certain amount of labeled data of tasks. Thus, these methods cannot scale to novel tasks with no labeled data. To the best of our knowledge, this is the *first* such work that involves regressing model parameters of novel tasks without using any ground truth information for the task.

Learning with Weak Supervision Our methodology exploits task correlation as a kind of weak supervision coming from the "wisdom-of-crowd". Recently, [38] uses weak supervision in the form of labeling functions inside a generative adversarial network and learns a joint distribution of labeled image. In topic modeling, [2] uses heuristics to build a hierarchical topic model. Methods such as [42, 49] use a fixed number of user-defined labeling functions to generate labels of an unlabeled images. Such methods use Maximum Likelihood Estimation and factor graphs to integrate results of each labeling functions. Bach et al. [4] exploit the use of distant supervision signals as weak supervision in the structure learning framework. MeTaL [43] extends [42] by identifying multiple sub-tasks that follow a hierarchy and then provides labeling functions for sub-tasks. In the context of zero-shot task generalization, we found work in reinforcement learning (RL) [37] where an agent is assigned to perform on unseen instructions, or instructions with complex and longer sequences. The agent utilizes its training knowledge as a weak supervision and try to maximize its reward in an unseen environment. However, we note that the interpretation of task, as well as the objective in [37] are very different from our proposed work.

13.3   Methodology

In this work, we propose a meta-learning algorithm that learns from the model parameters of known tasks (tasks with ground truth) and regresses high-performing model parameters of novel zero-shot tasks (tasks for which we do not have an access to the ground truth). Our meta-learner is assigned to accomplish K vision tasks, i.e. $\mathcal{T} = \{\tau_1, \tau_2, \cdots, \tau_m\}$. Our methodology assumes the model parameters of each task lie on a meta-manifold $\mathcal{M}_\theta$. Of K tasks, we refer to the first m tasks as known tasks, i.e. $\{\tau_1, \tau_2, \cdots, \tau_m\}$, since we have access to ground truths of these tasks. Consequently, their model parameters $\{(\theta_i) : i = 1, 2, \cdots, m\}$ and position on the meta-manifold $\mathcal{M}_\theta$ are known. The rest of the tasks, $\{\tau_{(m+1)}, \cdots, \tau_K\}$, are referred to as *zero-shot tasks* for convenience, as we have no access to ground truth for this set of tasks. We seek to model a meta-learner $\mathcal{F}(.)$ that learns from the model parameters of known tasks, and then regresses the model parameters of zero-shot tasks, i.e.:

$$\mathcal{F}(\theta_{\tau_1}, \theta_{\tau_2}, \cdots, \theta_{\tau_m}) = \theta_{\tau_j}, \qquad j = m + 1, \cdots, K \qquad (13.1)$$

However, building a meta-learner $\mathcal{F}(.)$ *only* from the knowledge of known tasks is not plausible, since the meta-learner is free to learn any co-ordinate on the meta-manifold, instead of converging to the desired point. To build a meta-learner that works in practice, we enable $\mathcal{F}(.)$ to leverage the knowledge of inter-task relationships. We introduce a task correlation matrix, Γ, as an additional input to the meta-learning function $\mathcal{F}(.)$, along with the known task model parameters. The task correlation matrix is an asymmetric matrix, where each entry $\gamma_{i,j} \in \Gamma$ represents the degree of task transfer between two tasks $\tau_i, \tau_j \in \mathcal{T}$. In this way, task correlation matrix Γ acts as a conditional to the meta-learning network $\mathcal{F}(.)$. Hence, our Eq. 13.1 now becomes:

$$\mathcal{F}(\theta_{\tau_1}, \theta_{\tau_2}, \cdots, \theta_{\tau_m}, \Gamma) = \theta_{\tau_j}, \qquad j = m + 1, \cdots, K \qquad (13.2)$$

The meta-learner $\mathcal{F}(.)$ itself is a network with parameters W, as in Fig. 13.1. We use the following objective function/loss to obtain the optimum value of W for the $\mathcal{F}(.)$:

$$\min_{W} \sum_{i=1}^{m} ||\mathcal{F}\big((\theta_{\tau_1}, \gamma_{1,i}), \cdots, (\theta_{\tau_m}, \gamma_{m,i})\big); W) - \theta_{\tau_i}^*||^2 \qquad (13.3)$$

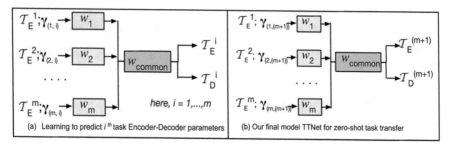

Fig. 13.1 Overview of our work. Figure (**a**) represents the training phase of TTNet, where it learns a correspondence between task correlation $\gamma_{(i,j)}$ of τ^i and τ^j, and the physical distance of τ^i and τ^j on the meta-manifold, given the encoder-decoder parameters of m-known tasks and the task correlation, the proposed TTNet gives encoder-decoder parameters of known tasks; (**b**) Once TTNet learns a correspondence between task correlation and the physical distance on the manifold, it can regress zero-shot task parameters given the task correlation of zero-shot task and m-known tasks. Please see Sect. 13.3 for more details (*Image Credit: [39]*)

Without loss of generality, all task model parameters can be assumed to have an encoder-decoder network structure, as in [54] (although our methodology is not necessarily constrained by such an assumption). And, by doing so, any task model parameter θ_{τ_i} can be thought to be comprised of an encoder network $\theta_{E_{\tau_i}}$ and a decoder network $\theta_{D_{\tau_i}}$. We note that in this work, the encoder networks of source tasks by themselves are considered sufficient to regress the encoder and decoder networks of zero-shot tasks $\{\tau_{(m+1)}, \cdots, \tau_K\}$. We hence rewrite Eq. 13.3 as follows:

$$\min_{W} \sum_{i=1}^{m} \left\| \mathcal{F}\big((\theta_{E_{\tau_1}}, \gamma_{1,i}), \cdots, (\theta_{E_{\tau_m}}, \gamma_{m,i}); W\big) - (\theta_{E_{\tau_i}}^*, \theta_{D_{\tau_i}}^*) \right\|^2 \qquad (13.4)$$

In this equation, $\theta_{E_{\tau_i}}^*$ and $\theta_{D_{\tau_i}}^*$ are the encoder and decoder parameters for the learned model of a known task, $\tau_i \in \mathcal{T}$. To the end of this section and later in Sect. 13.4, we provide more details of getting $\theta_{E_{\tau_i}}^*$, $\theta_{D_{\tau_i}}^*$ and model parameters of known tasks.

Minimizing the loss of the above loss function on the meta-manifold $\mathcal{M}_\theta$ alone may not be sufficient. It is important that the regressed encoder-decoder parameters perform well (low loss) on the actual task. To accomplish this, a *data model consistency loss* is added to Eq. 13.4. Let $\mathcal{D}_{\theta_{D_{\tau_i}}}(.)$ denote the data decoder parametrized by $\theta_{D_{\tau_i}}$, and $\mathcal{E}_{\theta_{E_{\tau_i}}}(.)$ denote the data encoder parameterized by $\theta_{E_{\tau_i}}$. Given an input image X, the regressed encoder-decoder model parameters should minimize the loss with respect to the ground truth y, i.e.:

$$\min_{W} \sum_{i=1}^{m} \left|\left| \mathcal{F}\big((\theta_{E_{\tau_1}}, \gamma_{1,i}), \cdots, (\theta_{E_{\tau_m}}, \gamma_{m,i}); W\big) - (\theta^*_{E_{\tau_i}}, \theta^*_{D_{\tau_i}}) \right|\right|^2$$

$$+ \lambda \sum_{\substack{x \in X_{\tau_i} \\ y \in \mathbf{y}_{\tau_i}}} \mathcal{L} \left(\mathcal{D}_{\tilde{\theta}_{D_{\tau_i}}} (\mathcal{E}_{\tilde{\theta}_{E_{\tau_i}}}(x)), y \right) \tag{13.5}$$

The Meta-network Following our objective function in Eq. 13.5, $\mathcal{F}(.)$ takes input from m number of known tasks. To accommodate encoder parameters of m known tasks, $\mathcal{F}(.)$ has m branches initially. Each of these m branches are parametrized by $\{W_1, W_2, \cdots, W_m\}$ respectively. Subsequently, the W_is are concatenated and fed into a W_{common} block that regresses the encoder and decoder model parameters. Driven by the intuition discussed in [52], we observed that the initial m branches parametrized by $W_i|_{i=1}^{m}$ encodes learning from individual tasks into a common representation space, and later layers parametrized by W_{common} learn to correlate the relationships between the task representations resulting in generalization to the zero-shot tasks. More specific details of $\{W_1, W_2, \cdots, W_m\}$, W_{common}, and all layers of $\mathcal{F}(.)$ are discussed in Sect. 13.4.

Learning the Task Correlation Matrix Γ Our methodology uses a task correlation matrix Γ to learn the relationships between tasks. This, as stated earlier, is obtained from crowdsourcing, although our proposed framework is not limited by this and can admit any other notion of task correlation. We have shown results where task correlation matrix, i.e. Γ_{TN}, is obtained through other methods such as [54] later in this chapter in Sect. 13.4.

Input The meta network is trained on a batch of model parameters of each known task $\tau_1, \cdots, \tau_m$. We closely follow the procedure of [51] to get p model parameters for each task. We assume that the size of the dataset B_j is S_j having instances $(X_{\tau_j}, \mathbf{y}_{\tau_j})$ corresponding to task τ_j. To get *one* encoder-decoder model parameter set, θ_{τ_j}, for this known task τ_j, we follow the following procedure: a base learner, i.e. encoder-decoder, defined by $\mathcal{D}(\mathcal{E}(x; \theta_{E_{\tau_j}}); \theta_{D_{\tau_j}})$ is trained on a subset b_j^k of the dataset B_j of task τ_j (of size s, $s \ll S_j$) with an appropriate loss function for the known task (mean-square error, cross-entropy or the like, based on the task). As a result, one instance of encoder-decoder model parameters, $\Theta_{\tau_j}^1 = \{\theta_{E_{\tau_j}}^1, \theta_{D_{\tau_j}}^1\}$, of task τ_j is obtained. Similarly, p subsets of labeled data, i.e. $\{b_j^1, b_j^2, \cdots, b_j^p\}$, of task τ_j are obtained using a sampling-with-replacement strategy from the dataset B_j. Following this, a set of p model parameters (one for each of p subsets sampled), i.e. $\Theta_{\tau_j} = \{\Theta_{\tau_j}^1, \cdots, \Theta_{\tau_j}^p\}$, for task τ_j is obtained. A similar process is followed to obtain p "optimal" model parameters for each known task $\{\Theta_{\tau_1}, \cdots, \Theta_{\tau_m}\}$. These model parameters (a total of $p \times m$ across all m known tasks) serve as the input to our meta network $\mathcal{F}(.)$.

On the other hand, we get the optimal encoder-decoder model parameters by training another base learner defined by $\mathcal{D}(\mathcal{E}(x; \theta_{E_{\tau_j}}); \theta_{D_{\tau_j}})$, trained on the full dataset B_j of task τ_j. We denote such encoder-decoder model parameters as $\Theta_{\tau_i}^* =$

$\{\theta^*_{E_{\tau_j}}, \theta^*_{D_{\tau_j}}\}$, and these serve as ground truth for the meta-learning function $\mathcal{F}(.)$ in Eqs. 13.3 and 13.5.

The Training of Meta-learner To learn W, the $\mathcal{F}(.)$ is trained on the objective function described in Eq. 13.5. Given the m known tasks model parameters $\{\theta_{\tau_1}, \theta_{\tau_2}, \cdots, \theta_{\tau_m}\}$ (obtained as above) and Γ, $\mathcal{F}(.)$ is trained in two modes: (1) *Self mode*, and (2) *Transfer mode*. Given a known task τ_i, the network $\mathcal{F}(.)$ updates weights in *only* the ith branch W_i and the W_{common} in the *self mode*, skipping parameters from all other $(m-1)$ branches. While, in *transfer mode*, the network $\mathcal{F}(.)$ updates weights of all other $(m-1)$ branches, i.e. $W_{\neg i}$ implying all $W_{j \neq i}, j = 1, \cdots, m$ and the W_{common}, but now skipping parameters from the ith branch. In *self mode* the $\mathcal{F}(.)$ operates in an autoencoder like manner that enforces regressed model parameters, θ_{τ_i}, to be as close to the actual model parameters $\theta_{\tau_i^*}$ (learned from the ground truth). While, in *transfer mode*, the $\mathcal{F}(.)$ learns a transfer of tasks from a set of source tasks (other than τ_i) and maps model parameters to a position close to the actual model parameters θ_{τ^*} on the meta manifold $\mathcal{M}_\theta$. The meta learner $\mathcal{F}(.)$ operates in *transfer mode* at inference time (i.e. for zero-shot task transfer).

Regressing Zero-Shot Parameters Given the optimal parameters W^* (learned after following Algorithm 5), the encoder model parameters of known tasks and Γ, the $\mathcal{F}(.)$ regresses encoder and decoder model parameters of zero-shot tasks, i.e. $\mathcal{F}_{W^*}\big((\theta_{E\tau_1}, \gamma_{1,j}), \cdots, (\theta_{E\tau_m}, \gamma_{m,j})\big)$ for all $j = (m+1), \cdots, T$. Additionally, we note that the task ordering does not make any noticeable difference in the process of implementing Algorithm 5.

Algorithm 5 Training our meta network, TTNet

Input: Number of epochs - Num_Epochs; Number of iterations needed for self mode/transfer mode - k; Optimal model parameters $\{\Theta^*_{\tau_1}, \cdots, \Theta^*_{\tau_m}\}$ of known tasks; Task correlation matrix $\Gamma = \{\gamma_{ij}\}$ ($m \times T$ matrix)
Output: Trained TTNet model, $\mathcal{F}(.)$
1: **for** Num_Epochs **do**
2:    **for** $j = 1, \cdots, m$ **do**
3:      **for** k steps **do**   ▷ Self mode
4:        Update weights W_i, W_{common} of $\mathcal{F}(.)$ by optimizing:
$$\Big\|\mathcal{F}\big((\theta_{E_{\tau_1}}, \gamma_{1,i}), \cdots, (\theta_{E_{\tau_1}}, \gamma_{m,i}); W_i, W_{common}\big) - (\theta^*_{E_{\tau_i}}, \theta^*_{D_{\tau_i}})\Big\|^2$$
$$+\lambda \sum_{\substack{x \in X_{\tau_i} \\ y \in \mathbf{y}_{\tau_i}}} \mathcal{L}\ \big(\mathcal{D}_{\tilde{\theta}_{D\tau_i}}(\mathcal{E}_{\tilde{\theta}_{E\tau_i}}(x)), y\big)$$
5:      **end for**
6:      **for** k steps **do**   ▷ Transfer mode
7:        Update weights $W_{\neg i}$, W_{common} of $\mathcal{F}(.)$ by optimizing:
$$\Big\|\mathcal{F}\big((\theta_{E_{\tau_1}}, \gamma_{1,i}), \cdots, (\theta_{E_{\tau_1}}, \gamma_{m,i}); W_{\neg i}, W_{common}\big) - (\theta^*_{E_{\tau_i}}, \theta^*_{D_{\tau_i}})\Big\|^2$$
$$+\lambda \sum_{\substack{x \in X_{\tau_i} \\ y \in \mathbf{y}_{\tau_i}}} \mathcal{L}\ \big(\mathcal{D}_{\tilde{\theta}_{D\tau_i}}(\mathcal{E}_{\tilde{\theta}_{E\tau_i}}(x)), y\big)$$
8:      **end for**
9:    **end for**
10: **end for**

13.4   Experimental Results

To evaluate our proposed framework, we consider the 26 vision tasks defined in [54]. For our initial experiments, we considered four vision tasks as zero-shot: surface normal, depth estimation, room layout, and camera-pose estimation, out of 26 vision tasks. These zero-shot tasks were chosen based on the question of how expensive the data annotation may be, and the complexity associated with the learning process using a deep network. Three zero-shot tasks, viz. surface normal estimation, depth estimation and room layout estimation, require expensive sensors to get ground truth from RGB data. The fourth, camera pose estimation, task involves estimating six degrees-of-freedom to collect ground truth, and is generally considered an expensive task. Initially, we consider three different TTNets to accomplish them; (1) $TTNet_6$, (2) $TTNet_{10}$, and (3) $TTNet_{20}$—which consider 6, 10 and 20 vision tasks as known source tasks respectively, and study the performance of TTNets on the four abovementioned zero-shot tasks. The fourth TTNet is similar to $TTNet_{20}$ with 20 source tasks, but the regressed parameters are finetuned on a small amount, (20%), of data for the zero-shot tasks—thus providing a low level of supervision and is hence called $TTNet_{LS}$. We later study the performance by varying source and zero-shot tasks and report the results in Sect. 13.5.

13.4.1   Description of Dataset

In this work, we consider the publicly available Taskonomy dataset [54]. Taskonomy identifies 26 vision tasks and provides ground truth for all of them given RGB images. All results are shown in Sect. 13.4 considering 120K images for training, 16K images for validation, and, 17K images for test, as described in [54].

13.4.2   Implementation Details

Network Architecture   All source tasks encoder networks use a fully convolutional ResNet-50 model without pooling. The decoder structures depend on the source task, for example: (1) *pixel-to-pixel tasks* such as normal estimation, each have decoder comprised of 15 fully convolutional layers; and (2) *low-dimensional tasks* such as vanishing points, the decoder consists of 2–3 fully connected (FC) layers. To get one input sample for TTNet of task τ_j, i.e. $\Theta_{\tau_j}^1 = \{\theta_{E\tau_j}^1, \theta_{D\tau_j}^1\}$ (described in Sect. 13.3), the encoder-decoder network $\mathcal{D}(\mathcal{E}(x; \theta_{E\tau_j}); \theta_{D\tau_j})$ is trained on 1k data points sampled (with replacement) from the Taskonomy dataset. Following this, we create 5000 samples of the model parameters for each task τ_j. We get the ground truth encoder-decoder network parameters of task τ_j, i.e. $\Theta_{\tau_j}^* = \{\theta_{E\tau_j}^*, \theta_{D\tau_j}^*\}$ by training $\mathcal{D}(\mathcal{E}(x; \theta_{E\tau_j}); \theta_{D\tau_j})$ on the entire Taskonomy dataset. All these data networks were trained with mini-batch Stochastic Gradient Descent (SGD) using a batch size of 32, learning rate of 0.001, momentum factor of $(0.5, 0.99)$ and Adam as an optimizer.

TTNet As described in Sect. 13.3, the TTNet has m branches, depending on the model under consideration (TTNet$_m$: $m \in \{6, 10, 20\}$). Each of such initial branches among m branches are comprised of 15 fully convolutional (FCONV) layers followed by 14 FC layers. All of them are then channel-wise combined to form a common layer, that is followed by 15 FCONV layers. The overall architecture is shown in Fig. 13.1. The TTNet is trained with mini-batch SGD, batch size of 32, learning rate of 0.0001, momentum factor of $(0.5, 0.99)$ and Adam as an optimizer.

Task Correlation For each pair of task $\tau_i, \tau_j \in \mathcal{T}$, experts give the response of task correlation on a scale +2 (strong relation), +1 (mild relation), 0 (abstain), −1 (no relation), while +3 is reserved for self-relation. We take response from 30 annotators for each pair of tasks $\tau_i, \tau_j \in \mathcal{T}$ (both known and zero-shot sets of tasks). The crowd votes are aggregated using the well-known Dawid-skene (DS) algorithm [14], and the task correlation matrix is shown in Fig. 13.2.

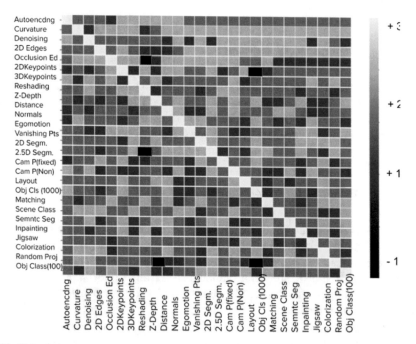

Fig. 13.2 Task correlation matrix. (Please zoom image to see details) We get the task correlation matrix Γ after receiving votes from 30 annotators. We use this Γ to build our meta-learner TTNet (*Image Credit: [39]*)

13.4.3 Comparison with State-of-the-Art Models

To study the effectiveness of our proposed TTNet, we report our empirical analysis both qualitatively and quantitatively on each of the four identified zero-shot tasks. We use $TTNet_6$ with a set of known tasks: autoencoding, denoising, 2D edges, occlusion edges, vanishing points and colorization, and all the results are presented using the correlation matrix Γ in Fig. 13.2.

13.4.4 Qualitative Results

In this section, we show qualitative results evaluating the performance of our model. For each of the zero shot tasks, we identify a separate set of state-of-the-art methods, and compare our proposed method with them and report them in the subsequent sections.

Surface Normal Surface normal estimation from a monocular RGB image is a challenging task in computer vision, and we try to accomplish this using the $TTNet_6$, $TTNet_{10}$, $TTNet_{20}$, and $TTNet_{LS}$ models. We identify state-of-the-art methods for this task: Multi-scale CNN (MC) [16], Deep3D (D3D) [50], Deep Network for surface normal estimation (DD) [50], SkipNet (Skip) [6], GeoNet (Geo) [41] and Taskonomy (TN) [54]. We compare TTNet against all of the above methods, and the results are shown in Fig. 13.3a. In the figure, the red boxes correspond to our models: $TTNet_6$, $TTNet_{10}$, $TTNet_{20}$, and $TTNet_{LS}$. With a increased number of source tasks, our TTNet shows improved results. For example, finer details (see edges of chandelier) are captured by $TTNet_{LS}$, which is not visible in other results.

Room Layout Estimation TTNet results are compared against: Volumetric (VM) [19], Edge-map (EM) [55], LayoutNet (LN) [56], RoomNet (RN) [27], and Taskonomy (TN) [54]. In Fig. 13.3b, green boxes represent TTNet results. Similar to [27] (which is a SegNet [5]-like encoder-decoder network) our encoder-decoder networks predict room layout keypoints and the corresponding edges and joins them with straight lines. To show the robustness on complex scenes, we report images with: (1) lot of occlusions, and (2) multiple edges such as roof-top, door, etc. in Fig. 13.3b (top and bottom respectively).

Depth Estimation Z-buffer depth estimation is the relative depth measurement from the camera plane. The encoder-decoder network regressed by TTNet computes depth from a single RGB image. The results of different TTNets are compared against state-of-the-art methods: FDA [28], Taskonomy [54], and GeoNet [41] as shown in Fig. 13.3c. TTNet results are shown in red bounding boxes, as before. To visualize the relative depth in Fig. 13.3c, the output is shown as a grayscale image (intensity values ranging from [0–255]), where near objects get intensity values closer to 0 (black) and the distant objects get intensity values closer to 255 (white). It can be observed from Fig. 13.3c that $TTNet_{10}$ outperforms [54]; and $TTNet_{20}$ and $TTNet_{LS}$ outperform all other methods studied.

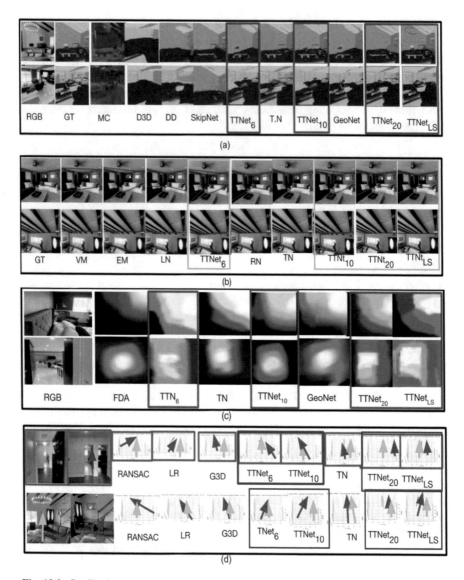

Fig. 13.3 Qualitative comparison (Best viewed in color): TTNet models compared against other state-of-the-art models, see Sect. 13.4.4 for details. (**a**) *Surface Normal Estimation:* Red boxes indicate results of our TTNet models; (**b**) *Room Layout:* Red edges indicate the predicted room edges; green boxes indicate our TTNet model results; (**c**) *Depth Estimation:* Red bounding boxes show our results; (**d**) *Camera Pose Estimation:* First image is the reference frame of the camera, i.e. green arrow. The second image, with red arrow, is taken after a geometric translation w.r.t first image. Blue rectangles show our results (*Image Credit: [39]*)

Table 13.1 Surface normal estimation

Method	Mean (↓)	Medn (↓)	RMSE (↓)	11.25° (↑)	22.5° (↑)	30° (↑)
MC[16]	30.30	35.30	–	30.29	57.17	68.29
DD [50]	25.71	20.81	31.01	38.12	59.18	67.21
DD[50]	21.10	15.61	–	44.39	64.48	66.21
Skip [6]	20.21	12.19	28.20	47.90	70.00	78.23
TN[54]	19.90	11.93	23.13	48.03	70.02	78.88
$TTNet_6$	19.22	12.01	26.13	48.02	71.11	78.29
Geo [41]	19.00	11.80	26.90	48.04	72.27	79.68
$TTNet_{10}$	19.81	11.09	22.37	48.83	71.61	79.00
$TTNet_{20}$	19.27	11.91	26.44	48.81	71.97	79.72
$TTNet_{LS}$	**15.10**	**9.29**	**24.31**	**56.11**	**75.19**	**84.71**

Mean, median and RMSE refer to the difference between the model's predicted surface normal and ground truth surface normal (a lower value is better, ↓). Other 3 are the number of pixels within 11.25°, 22.5° and 30° thresholds within ground truth's predicted pixels (a higher number is better, ↑). – indicates those values cannot be obtained by the corresponding method

Camera Pose Estimation For fixated camera pose estimation, all TTNet models are compared against state-of-the-art methods: RANSAC [15], Latent RANSAC [24], Generic3D pose [53] and Taskonomy [54]. We note that our zero-shot regressed encoder-decoder network for camera pose estimation is similar to the encoder-decoder based camera pose estimation method, called Hourglass Network [33]. In our framework, the encoder network takes two inputs of the same view captured from two different geometric points in a sequential manner. A fixed camera pose estimation model predicts any five of the 6-degrees of freedom: yaw, pitch, roll and x,y,z translation. We show our results in Fig. 13.3d: first image is the reference frame of the camera (green arrow), and second image (red arrow) is taken after a geometric transformation.

13.4.5 Quantitative Results

Surface Normal Estimation We evaluated our method based on the evaluation criteria described in [6, 41]. The results are presented in Table 13.1. Our $TTNet_6$ is comparable to state-of-the-art Taskonomy [54] and GeoNet [41]. Our $TTNet_{10}$, $TTNet_{20}$, and $TTNet_{LS}$ outperforms all state-of-the-art models.

Room Layout Estimation Evaluation criteria are: (1) *Keypoint error:* a global measurement averaged on Euclidean distance between model's predicted keypoint and the ground truth; and (2) *Pixel error:* a local measurement that estimates pixelwise error between the predicted surface labels and ground truth labels. Table 13.2 presents the results.

Table 13.2 Room layout estimation

Metd	VM [19]	EM [55]	LN [56]	TTNet$_6$	RN [27]	TN [54]	TTNet$_{10}$	TTNet$_{20}$	TTNet$_{LS}$
Key.	15.5	11.2	7.64	7.5	6.3	6.2	6.00	5.82	**5.52**
Pixel	24.3	16.7	10.6	8.1	8.0	8.0	7.72	7.10	**6.81**

Both TTNet$_{20}$ and TTNet$_{LS}$ outperformed state-of-the-art models on keypoint (Key. in table) and pixel error (Pixel in table)

Table 13.3 Depth estimation

Method	RMSE(lin)	RMSE(log)	ARD	SRD
FDA [28]	0.877	0.283	0.214	0.204
TTNet$_6$	0.745	0.262	0.220	0.210
TN [54]	0.591	0.231	0.242	0.206
TTNet$_{10}$	0.575	0.172	0.236	0.179
Geonet [41]	0.591	0.205	0.149	0.118
TTNet$_{20}$	0.597	0.204	0.140	0.106
TTNet$_{LS}$	**0.572**	**0.193**	**0.139**	**0.096**

TTNet$_{20}$ and TTNet$_{LS}$ outperform all other methods studied

Table 13.4 Camera pose estimation (fixed)

Method	RANSAC [52]	LR [24]	G3D [53]	TN [54]
TTNet$_6$	88%	81%	72%	64%
TTNet$_{10}$	90%	82%	79%	82%
TTNet$_{20}$	90%	82%	92%	80%
TTNet$_{LS}$	**96%**	**88%**	**96%**	**87%**

We have considered *win rate* (%) on angular error. Columns are state-of-the-art methods and rows are our four TTNet models

Depth Estimation Following [28], we use the evaluation criteria: RMSE $(\text{lin}) = \frac{1}{N}(\sum_X (d_X - d_X^*)^2)^{\frac{1}{2}}$; $\text{RMSE(log)} = \frac{1}{N}(\sum_X (\log d_X - \log d_X^*)^2)^{\frac{1}{2}}$; Absolute relative distance $= \frac{1}{N}\sum_X \frac{|d_X - d_X^*|}{d_X}$; Squared absolute distance $= \frac{1}{N}\sum_X \left(\frac{|d_X - d_X^*|}{d_X}\right)^2$. d_X^* is ground truth depth, d_X is estimated depth, and N is the total number of pixels in all images in the test set (Table 13.3).

Camera Pose Estimation We adopted the *win rate* (%) evaluation criteria [54] that counts the proportion of images for which a baseline is outperformed. Table 13.4 shows the win rate of TTNet models on angular error with respect to state-of-the-art models: RANSAC [52], LRANSAC [24], G3D and Taskonomy [54].

13.5 Discussion and Analysis

Relevance of Source Tasks for Regressing Zero-Shot Tasks To understand the effect of source tasks on regressing the zero-shot tasks, we follow the GO-MTL approach [25] and compute the task basis of the parameters of each source and zero-shot task. Optimal model parameters of each source task τ_i are mapped to a

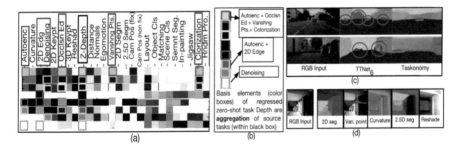

Fig. 13.4 (Please zoom to see more detail) (**a**) **Task basis analysis:** Task basis elements of Zero-shot task, i.e. z-depth (black box), is supported by six known tasks (blue box), which is found using GO-MTL [25]; (**b**) Zero-shot task basis is comprised of the elements of know tasks basis elements; (**c**) **Surface normal estimation on Cityscapes:** Red circles highlight details (car, tree, human) by TTNet; (**d**) **Different choice zero shot tasks:** 2D segmentation, Vanishing point estimation, Curvature estimation, 2.5D segmentation and reshading, other than Sect. 13.4 (*Image Credit: [39]*)

lower dimensional space R, i.e. $S : \tau_i \rightarrow R_i$ followed by a task basis analysis using GO-MTL. We show the result in Fig. 13.4a. A source task is considered to be important if it contributes a good number of task basis elements on building task basis of zero-shot tasks. For example, the source task "autoencoding" shares four basis elements with zero-shot task "depth estimation" and hence is an important source task while regressing "depth estimation" model parameters.

Why Predicted Zero-Shot Task Parameters Perform Better than Supervised Training? In continuation to the above discussion, the zero-shot task basis elements are composed of latent elements from several source tasks and provide more information. For example, in Fig. 13.4b, we can see that the basis vector of zero-shot task "Z-depth" is well-supported by latent elements of source tasks. Our setting hence shows results where learning from similar tasks provides good performance than learning from the specific task alone in a purely supervised setting.

Ablation Study on Different Number of Annotators All the previous evaluations were performed by aggregating votes of 30 experts (on the estimates of task correlation). We study the robustness of our method by varying the number of experts, M_i, where $i \in \{3, 10, 20, 30, 35, 40, 45, 50\}$. We report the *win rate (%)* of the pose estimation task using TTNet$_6$ with a varied number of experts. We found the results in all of these cases to be promising, and observed that our studies did not need beyond $i = 30$ number of annotators to get our best results, and we hence report the same across our experiments.

Transfer Learning from Zero-Shot Task to a Known Task We study the question of—how capable are the predicted zero-shot model parameters for transferring to a known task? To answer this, we take the regressed zero-shot encoder-decoder parameters and finetune the decoder to a target known task while fixing the encoder parameters, following the procedure described in [54]. Table 13.5 shows quantitative

Table 13.5 Zero-shot to known task transfer

| | TTNet$_6$ | | | | | | Taskonomy | | | | | |
| | Wang | | Zamir | | Full Sup | | Wang | | Zamir | | Full Sup | |
Model	N	L	N	L	N	L	N	L	N	L	N	L
Depth	85	**87**	81	**97**	**67**	42	**98**	85	**92**	88	60	**46**
2.5 D	**88**	75	**75**	81	**89**	35	**88**	77	73	88	85	39
Curvature	**84**	87	**91**	58	**86**	47	78	**89**	88	**78**	60	**50**

We consider the autoencoder-decoder parameters for a zero-shot task learned through our method, and finetune the decoder (fixing the encoder) to a target known task, following the procedure in [54]. Source tasks (zero-shot) are surface normal (N), and, room layout (L). Target tasks are depth, 2.5D segmentation and curvature. *Win rates* (%) of task transfer with respect to self-supervised methods, such as, Wang et al. [50], Zamir et al. [53] as well as fully supervised setting are shown (all values are in %), with bold face numerals denoting winning entries

results where the zero-shot tasks, surface normal estimation and room layout estimation, are used for transfer learning to target tasks: depth estimation, 2.5D segmentation and curvature (all known tasks). *Win rates* (in %) of transfer learning with respect to the fully supervised method and self-supervised methods: Wang et al. [51] and G3D [53] are shown in Table 13.5. We observe very encouraging results in the table.

Efficacy on Other Datasets Given the task correlation and the learned model parameters of source tasks, it is worthwhile to ask how generalizable the proposed method is to other visual domains, which are not necessarily as task-rich as the Taskonomy dataset. We examined this question by finetuning TTNet on the Cityscapes dataset [10] and COCO-Stuff dataset [7]. We choose surface normal estimation as a zero-shot task on the Cityscapes dataset and try our proposed TTNet and [54]. The results are shown in Fig. 13.4c—we clearly observe that TTNet captures more detail. Further, the TTNet$_6$ model was finetuned on the segmentation model parameters trained on Cityscapes data. We compared our method against [54] and note that despite having no supervision, the proposed method outperformed [54] on four zero-shot tasks: Surface normal, depth, 2D edge and 3D keypoint on the Cityscapes dataset. It is evident from Fig. 13.5 that our model seems to capture more detail. On the COCO-Stuff dataset, object detection is considered as a zero-shot task. TTNet is finetuned to perform object detection and the result is shown in Fig. 13.6. In both cases, our proposed method performs fairly well.

Optimal Number of Known Tasks To what extent changing: (1) the number of source tasks, and (2) the choice of source tasks used to regress zero-shot task model parameters, is a legitimate question to ponder. Since an exhaustive study across all possible combinations of source task is not feasible, we attempt to answer using a set of six different TTNet models: TTNet$_4$, TTNet$_6$, TTNet$_{10}$, TTNet$_{15}$, TTNet$_{18}$ and TTNet$_{20}$ (the subscript implies number of source tasks). In Fig. 13.7, we show a study with different number and combination of source tasks, and report the *win rates* (in %) against the Taskonomy method [54], for each of the zero-shot tasks.

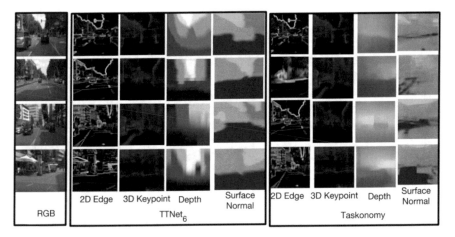

| | 2D Edge | 3D Keypoint | Depth | Surface Normal | 2D Edge | 3D Keypoint | Depth | Surface Normal |
| RGB | | | TTNet$_6$ | | | | Taskonomy | |

Fig. 13.5 Results on Cityscapes data. We finetuned TTNet$_6$ on the Cityscapes dataset [10], and the surface normal, depth, 2D edge and 3D keypoint results are reported using the model parameters learned by TTNet$_6$ (*Image Credit: [39]*)

Method	AP{50:95}	AP{50}	AP{75}	AP{sml}	AP{med}	AP{lrg}
CoupleNet	34.4	54.8	37.2	13.4	8.1	50.8
TTNet{6}	29.9	51.9	34.6	10.8	32.8	45
YOLOv2	21.6	44	19.2	5	22.4	35.5

Fig. 13.6 Object Detection using TTNet$_6$: TTNet$_6$ is finetuned on the COCO-stuff dataset (*Image Credit: [39]*)

Figure 13.7 shows that an increased number of source tasks improves performance of regressed zero-shot tasks. Despite having a small number of source tasks, and the fact that most of the source tasks are cheap for annotation (i.e. autoencoding, denoising, 2D edges, occlusion edges, vanishing points and colorization—all do not require significant annotation), TTNet$_6$ performs well. We also note that TTNet$_6$ is quite robust under the change of choice of source tasks.

Zero-Shot Task Transfer Under Alternate Task Correlation Matrix Thus far, we used the task correlation matrix Γ by aggregating the "wisdom of crowd". This section exploits an alternate approach of considering task graph obtained in [54] to build the task correlation matrix Γ. We refer to this new task correlation matrix as Γ_{TN} and show a qualitative comparison in Fig. 13.8. We note however that the task correlation matrix Γ_{TN} obtained through the work of [54] requires ground truth labeled data of all tasks *including* zero-shot tasks. It may hence not qualify as a proper replacement of the Γ in our framework, which is crowdsourced without any labeled data from zero-shot tasks.

	4			6			10			15		18		20
Win Rate (Camera Ps.(fixed)) (%)	71%	59%	52%	89%	89%	82%	85%	92%	89%	93%	85%	91%	94%	89%
Win Rate (Depth) (%)	71%	61%	79%	87%	86%	86%	87%	91%	81%	91%	81%	97%	93%	93%
Win Rate (Room Layout) (%)	62%	58%	79%	85%	86%	88%	84%	88%	83%	85%	87%	91%	91%	91%
Win Rate (Normal) (%)	79%	71%	75%	88%	87%	85%	85%	87%	88%	88%	89%	93%	95%	94%
Colorization	✗	✓	✓	✓	✓	✗	✓	✓	✓	✓	✓	✓	✓	✓
Random Projection	✗	✓	✗	✗	✗	✓	✓	✓	✗	✓	✓	✓	✓	✓
Jig-Saw Puzzle	✓	✓	✓	✗	✗	✓	✓	✓	✗	✓	✓	✓	✓	✓
Vanishing Point	✗	✗	✗	✓	✓	✗	✗	✓	✗	✓	✓	✓	✓	✓
Semantic Segmentation	✗	✗	✗	✗	✗	✗	✗	✓	✓	✓	✓	✓	✓	✓
2D Segmentation	✗	✗	✗	✗	✗	✗	✗	✗	✓	✓	✓	✗	✓	✓
2.5D Segmentation	✗	✗	✗	✗	✗	✗	✗	✗	✗	✗	✓	✓	✓	✓
Room Layout	✗	✗	✗	✗	✗	✗	✗	✗	✗	✗	✗	✗	✗	✗
Normals	✗	✗	✗	✗	✗	✗	✗	✗	✗	✗	✗	✗	✗	✗
Distance	✗	✗	✗	✗	✗	✗	✗	✗	✗	✗	✗	✓	✓	✓
Z-Depth	✗	✗	✗	✗	✗	✗	✗	✗	✗	✗	✗	✗	✗	✗
Reshading	✗	✗	✗	✗	✓	✗	✓	✗	✓	✗	✓	✓	✓	✓
Matching	✗	✗	✗	✗	✗	✗	✗	✗	✗	✗	✓	✓	✓	✓
Cam Pose (non-fixed)	✗	✗	✗	✗	✗	✗	✗	✗	✗	✗	✓	✓	✓	✓
3D Key Point	✗	✗	✗	✗	✓	✗	✗	✗	✗	✓	✗	✓	✗	✓
2D Key Point	✗	✗	✗	✗	✗	✗	✓	✗	✗	✗	✗	✓	✓	✓
Cam Pose (fixed)	✗	✗	✗	✗	✗	✗	✗	✗	✗	✗	✗	✗	✗	✗
Ego motion	✗	✗	✗	✗	✗	✗	✗	✗	✓	✓	✓	✓	✓	✓
Occlusion Edges	✗	✗	✗	✓	✗	✗	✗	✓	✗	✓	✗	✗	✗	✓
2D Edges	✓	✗	✗	✓	✗	✗	✗	✗	✗	✓	✓	✓	✓	✓
Denoising	✓	✓	✗	✓	✗	✗	✓	✓	✓	✓	✓	✓	✓	✓
Curvature	✗	✗	✗	✗	✗	✓	✓	✓	✓	✓	✓	✓	✓	✓
Scene Class	✗	✗	✗	✗	✓	✓	✓	✗	✓	✓	✗	✓	✓	✓
Object Class	✗	✗	✗	✗	✗	✓	✓	✓	✓	✓	✗	✓	✗	✗
Autoencoding	✓	✓	✓	✓	✓	✓	✓	✓	✓	✓	✓	✓	✓	✓
	4			6			10			15		18		20

Fig. 13.7 Ablation study of different TTNets **with different combinations of source tasks.** Column numerals, i.e. 4, 6, 10, 15, 18 and 20 (bottom row) indicate the number of source tasks that have been considered to train the model. A blue box (with ✓) indicates a source task. The first four rows indicate *win rates* (%) for each of the 4 zero-shot tasks considered against [54]. We observe that our methodology is fairly robust and that TTNet$_6$ gives a fairly good performance even with a low number of source tasks (*Image Credit: [39]*)

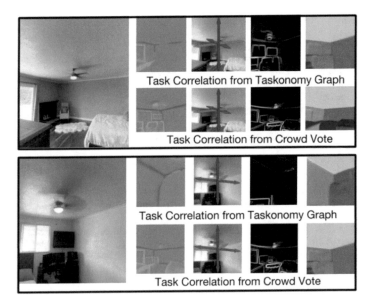

Fig. 13.8 Qualitative results of TTNet$_6$ **when task correlation matrix (Γ) is obtained from task graph computed in [54].** We studied considering the task graph computed in [54] (instead of crowd vote) to build the task correlation matrix Γ_{TN}. First column represents RGB image and, subsequent columns (from second to fourth columns) are zero-shot tasks: curvature, vanishing points, 2D key point and surface normal estimation (*Image Credit: [39]*)

13.6   Conclusion

In this chapter, we presented a meta-learner, TTNet, to regress model parameters of novel tasks for which no ground truth is available (*zero-shot tasks*). One could relate the inspiration for the proposed method to findings in cognitive science which show that several mammals adapt to an entirely novel task (depth measurement) by understanding and association with old learned tasks (self-motion, shoulder movements) without receiving an explicit supervision. TTNet regresses encoder-decoder model parameters of four zero-shot tasks: surface normal estimation, room layout estimation, depth estimation and camera pose estimation, although we also show the method is robust despite other choices of tasks. The results are compared with [54] and other state-of-the-art methods. Our future work involves learning task correlation from various sources and the corresponding results for zero-shot task transfer. In particular, negative transfer in task space is a particularly interesting direction of future work.

References

1. Agarwal, A., Gerber, S., Daume, H.: Learning multiple tasks using manifold regularization. In: Advances in Neural Information Processing Systems, pp. 46–54 (2010)
2. Alfonseca, E., Filippova, K., Delort, J.Y., Garrido, G.: Pattern learning for relation extraction with a hierarchical topic model. In: Proceedings of the 50th Annual Meeting of the Association for Computational Linguistics: Short Papers-Volume 2, pp. 54–59. Association for Computational Linguistics, Stroudsburg (2012)
3. Argyriou, A., Evgeniou, T., Pontil, M.: Convex multi-task feature learning. Mach. Learn. **73**(3), 243–272 (2008)
4. Bach, S.H., He, B., Ratner, A., Ré, C.: Learning the structure of generative models without labeled data (2017). Preprint. arXiv:1703.00854
5. Badrinarayanan, V., Kendall, A., Cipolla, R.: Segnet: a deep convolutional encoder-decoder architecture for image segmentation. IEEE Trans. Pattern Anal. Mach. Intell. **39**(12), 2481–2495 (2017)
6. Bansal, A., Russell, B., Gupta, A.: Marr revisited: 2d-3d alignment via surface normal prediction. In: Proceedings of the IEEE Conference on Computer Vision and Pattern Recognition, pp. 5965–5974 (2016)
7. Caesar, H., Uijlings, J., Ferrari, V.: Coco-stuff: thing and stuff classes in context. In: Proceedings of the IEEE Conference on Computer Vision and Pattern Recognition, pp. 1209–1218 (2018)
8. Chen, Q., Liu, Y., Wang, Z., Wassell, I., Chetty, K.: Re-weighted adversarial adaptation network for unsupervised domain adaptation. In: Proceedings of the IEEE Conference on Computer Vision and Pattern Recognition, pp. 7976–7985 (2018)
9. Cohen, H., Crammer, K.: Learning multiple tasks in parallel with a shared annotator. In: Advances in Neural Information Processing Systems, pp. 1170–1178 (2014)
10. Cordts, M., Omran, M., Ramos, S., Rehfeld, T., Enzweiler, M., Benenson, R., Franke, U., Roth, S., Schiele, B.: The cityscapes dataset for semantic urban scene understanding. In: Proceedings of the IEEE Conference on Computer Vision and Pattern Recognition (CVPR) (2016)
11. Csurka, G.: Domain adaptation for visual applications: a comprehensive survey (2017). Preprint. arXiv:1702.05374
12. Dai, W., Yang, Q., Xue, G.R., Yu, Y.: Boosting for transfer learning. In: Proceedings of the 24th International Conference on Machine Learning, pp. 193–200. ACM, New York (2007)
13. Daumé III, H.: Bayesian multitask learning with latent hierarchies (2009). Preprint. arXiv:0907.0783
14. Dawid, A.P., Skene, A.M.: Maximum likelihood estimation of observer error-rates using the em algorithm. Appl. Stat. **28**, 20–28 (1979)
15. Derpanis, K.G.: Overview of the RANSAC algorithm. Image Rochester NY **4**(1), 2–3 (2010)
16. Eigen, D., Fergus, R.: Predicting depth, surface normals and semantic labels with a common multi-scale convolutional architecture. In: Proceedings of the IEEE International Conference on Computer Vision, pp. 2650–2658 (2015)
17. Finn, C., Abbeel, P., Levine, S.: Model-agnostic meta-learning for fast adaptation of deep networks (2017). Preprint. arXiv:1703.03400
18. Ghifary, M., Kleijn, W.B., Zhang, M., Balduzzi, D., Li, W.: Deep reconstruction-classification networks for unsupervised domain adaptation. In: European Conference on Computer Vision, pp. 597–613. Springer, Cham (2016)
19. Gupta, A., Hebert, M., Kanade, T., Blei, D.M.: Estimating spatial layout of rooms using volumetric reasoning about objects and surfaces. In: Advances in Neural Information Processing Systems, pp. 1288–1296 (2010)
20. Ha, D., Dai, A., Le, Q.V.: Hypernetworks (2016). Preprint. arXiv:1609.09106
21. Held, R., Hein, A.: Movement-produced stimulation in the development of visually guided behavior. J. Comput. Physiol. Psychol. **56**(5), 872 (1963)

22. Kale, D., Liu, Y.: Accelerating active learning with transfer learning. In: 2013 IEEE 13th International Conference on Data Mining, pp. 1085–1090. IEEE, Piscataway (2013)
23. Kim, T., Yoon, J., Dia, O., Kim, S., Bengio, Y., Ahn, S.: Bayesian model-agnostic meta-learning (2018). Preprint. arXiv:1806.03836
24. Korman, S., Litman, R.: Latent RANSAC. In: Proceedings of the IEEE Conference on Computer Vision and Pattern Recognition, pp. 6693–6702 (2018)
25. Kumar, A., Daume III, H.: Learning task grouping and overlap in multi-task learning (2012). Preprint. arXiv:1206.6417
26. Lee, S.I., Chatalbashev, V., Vickrey, D., Koller, D.: Learning a meta-level prior for feature relevance from multiple related tasks. In: Proceedings of the 24th International Conference on Machine learning, pp. 489–496. ACM, New York (2007)
27. Lee, C.Y., Badrinarayanan, V., Malisiewicz, T., Rabinovich, A.: Roomnet: end-to-end room layout estimation. In: 2017 IEEE International Conference on Computer Vision (ICCV), pp. 4875–4884. IEEE, Piscataway (2017)
28. Lee, J.H., Heo, M., Kim, K.R., Kim, C.S.: Single-image depth estimation based on fourier domain analysis. In: Proceedings of the IEEE Conference on Computer Vision and Pattern Recognition, pp. 330–339 (2018)
29. Liu, M.Y., Tuzel, O.: Coupled generative adversarial networks. In: Advances in Neural Information Processing Systems, pp. 469–477 (2016)
30. Liu, S., Pan, S.J., Ho, Q.: Distributed multi-task relationship learning. In: Proceedings of the 23rd ACM SIGKDD International Conference on Knowledge Discovery and Data Mining, pp. 937–946. ACM, New York (2017)
31. Luo, Z., Zou, Y., Hoffman, J., Fei-Fei, L.F.: Label efficient learning of transferable representations across domains and tasks. In: Advances in Neural Information Processing Systems, pp. 165–177 (2017)
32. Madrid, J., Escalante, H.J., Morales, E.: Meta-learning of textual representations (2019). Preprint. arXiv:1906.08934
33. Melekhov, I., Ylioinas, J., Kannala, J., Rahtu, E.: Image-based localization using hourglass networks. In: Proceedings of the IEEE International Conference on Computer Vision, pp. 879–886 (2017)
34. Mihalkova, L., Huynh, T., Mooney, R.J.: Mapping and revising Markov logic networks for transfer learning. In: AAAI, vol. 7, pp. 608–614 (2007)
35. Mishra, N., Rohaninejad, M., Chen, X., Abbeel, P.: A simple neural attentive meta-learner (2017). Preprint. arXiv:1707.03141
36. Naik, D.K., Mammone, R.: Meta-neural networks that learn by learning. In: International Joint Conference on Neural Networks, 1992, IJCNN, vol. 1, pp. 437–442. IEEE, Piscataway (1992)
37. Oh, J., Singh, S., Lee, H., Kohli, P.: Zero-shot task generalization with multi-task deep reinforcement learning (2017). Preprint. arXiv:1706.05064
38. Pal, A., Balasubramanian, V.N.: Adversarial data programming: using gans to relax the bottleneck of curated labeled data. In: Proceedings of the IEEE Conference on Computer Vision and Pattern Recognition, pp. 1556–1565 (2018)
39. Pal, A., Balasubramanian, V.N.: Zero-shot task transfer. In: Proceedings of the IEEE Conference on Computer Vision and Pattern Recognition, pp. 2189–2198 (2019)
40. Pan, S.J., Yang, Q.: A survey on transfer learning. IEEE Trans. Knowl. Data Eng. **22**(10), 1345–1359 (2009)
41. Qi, X., Liao, R., Liu, Z., Urtasun, R., Jia, J.: Geonet: geometric neural network for joint depth and surface normal estimation. In: Proceedings of the IEEE Conference on Computer Vision and Pattern Recognition, pp. 283–291 (2018)
42. Ratner, A.J., De Sa, C.M., Wu, S., Selsam, D., Ré, C.: Data programming: creating large training sets, quickly. In: Advances in Neural Information Processing Systems, pp. 3567–3575 (2016)
43. Ratner, A., Hancock, B., Dunnmon, J., Sala, F., Pandey, S., Ré, C.: Training complex models with multi-task weak supervision (2018). Preprint. arXiv:1810.02840

44. Redmon, J., Divvala, S., Girshick, R., Farhadi, A.: You only look once: unified, real-time object detection. In: Proceedings of the IEEE Conference on Computer Vision and Pattern Recognition, pp. 779–788 (2016)
45. Saito, K., Watanabe, K., Ushiku, Y., Harada, T.: Maximum classifier discrepancy for unsupervised domain adaptation **3** (2017). Preprint. arXiv:1712.02560
46. Sener, O., Song, H.O., Saxena, A., Savarese, S.: Learning transferrable representations for unsupervised domain adaptation. In: Advances in Neural Information Processing Systems, pp. 2110–2118 (2016)
47. Smith, L., Gasser, M.: The development of embodied cognition: six lessons from babies. Artif. Life **11**(1–2), 13–29 (2005)
48. Thrun, S., Pratt, L.: Learning to learn. Springer Science & Business Media, New York (2012)
49. Varma, P., He, B., Iter, D., Xu, P., Yu, R., De Sa, C., Ré, C.: Socratic learning: augmenting generative models to incorporate latent subsets in training data (2016). Preprint. arXiv:1610.08123
50. Wang, X., Fouhey, D., Gupta, A.: Designing deep networks for surface normal estimation. In: Proceedings of the IEEE Conference on Computer Vision and Pattern Recognition, pp. 539–547 (2015)
51. Wang, Y.X., Ramanan, D., Hebert, M.: Learning to model the tail. In: NIPS (2017)
52. Yosinski, J., Clune, J., Bengio, Y., Lipson, H.: How transferable are features in deep neural networks? In: Advances in Neural Information Processing Systems, pp. 3320–3328 (2014)
53. Zamir, A.R., Wekel, T., Agrawal, P., Wei, C., Malik, J., Savarese, S.: Generic 3d representation via pose estimation and matching. In: European Conference on Computer Vision, pp. 535–553. Springer, Cham (2016)
54. Zamir, A.R., Sax, A., Shen, W., Guibas, L., Malik, J., Savarese, S.: Taskonomy: disentangling task transfer learning. In: Proceedings of the IEEE Conference on Computer Vision and Pattern Recognition, pp. 3712–3722 (2018)
55. Zhang, W., Zhang, W., Liu, K., Gu, J.: Learning to predict high-quality edge maps for room layout estimation. IEEE Trans. Multimedia **19**(5), 935–943 (2017)
56. Zou, C., Colburn, A., Shan, Q., Hoiem, D.: Layoutnet: reconstructing the 3d room layout from a single RGB image. In: Proceedings of the IEEE Conference on Computer Vision and Pattern Recognition, pp. 2051–2059 (2018)

Printed in the United States
by Baker & Taylor Publisher Services